THE STUDY OF ANCIENT TERRITORIES

CHERSONESOS & METAPONTO

2002 ANNUAL REPORT

INSTITUTE OF CLASSICAL ARCHAEOLOGY
THE UNIVERSITY OF TEXAS
AUSTIN

The Director and Staff
of
The Institute of Classical Archaeology
gratefully acknowledge the extraordinary support
for operations and projects

provided by

The Packard Humanities Institute

Special thanks also to:

Non-profit Organizations

The Bernard and Audré Rapoport Foundation
The Brown Foundation
Center for the Study of Ancient Territories (CSAT)
The Dumbarton Oaks Center for Byzantine Studies
The Global Heritage Foundation
The James R. Dougherty, Jr., Foundation
The Liss Foundation
The Marian and Speros Martel Foundation
The Rachael & Ben Vaughan Foundation
The Society of Antiquaries of London
The Trust for Mutual Understanding
The World Monuments Fund

Governmental Institutions

Embassy of Ukraine in the United States of America
Embassy of the United States of America in Ukraine
Ministero degli Affari Esteri (Direzione Generale per la Promozione e la Cooperazione Culturale)
Museo Archeologico Nazionale, Metaponto
National Aeronautics and Space Administration (NASA)
The National Preserve of Tauric Chersonesos, Ukraine
Regione Basilicata
Sevastopol City Administration
Soprintendenza Archeologica della Basilicata, Italy

Educational Institutions

Institute of Archaeology, Ukrainian National Academy
National Taras Shevchenko University of Kyiv
Taurian National University, Simferopol
The University of Kyiv Mohyla Academy
Università degli Studi di Lecce
- Amministrazione Centrale
- Dipartimento di Beni Culturali
- Prof. Oronzo Limone
- Prof. Mario Lombardo
- Prof. Linda Safran
- Prof. Chris Wickham

The University of Texas at Austin
- College of Liberal Arts
- Department of Classics
- Center for Russian, East European, and Eurasian Studies (CREES)
- Center for Space Research (CSR)
- School of Information
- Harry Ransom Humanities Research Center (HRC)

Individual Donors

Mr. and Mrs. Daniel Balz
Ms. Jane Bedichek
Col. Dayton Blanchard
Mrs. Mary Leaphart Carter
Dr. William L. Carter, M.D.
Ms. Judy Coker
Mr. Jack Englert, Jr.
Ms. Sandra Heinrichs
Mr. and Mrs. Marshall Johnson
Mr. and Mrs. Daniel Kruger
Mr. Warren Larson
Dr. and Mrs. Henry Liss
Ms. Mary Link Malone
Prof. Lucy Shoe Meritt
Mr. Bert Michelsen
Ms. Kathleen Monahan
Mr. and Mrs. John O'Connor
Dr. Maconda Brown O'Connor
Mr. Ralph S. O'Connor
Mr. Bernard Rapoport
Dr. and Mrs. David M. Renison
Mrs. Elizabeth H. Tracey
Mr. and Mrs. Ben F. Vaughan III
Ms. Nancy C. York

Contents

The Institute of Classical Archaeology

Mission

The Institute of Classical Archaeology (ICA) is an Organized Research Unit of the University of Texas at Austin. In order to study, document, and preserve sites, monuments, and artifacts of past life for present and future generations, we conduct archaeological fieldwork and training in ancient Greek rural territories on the Black Sea and the Mediterranean. We engage in interdisciplinary research and publication and provide technical assistance and expertise for cultural heritage management.

Research Focus

ICA's research focuses primarily on the *chorai*—or agricultural territories—that surrounded and supported ancient Greek colonial cities. Our two primary sites of research are the *chora* of Metaponto on the southern coast of Italy and the *chora* of Chersonesos on the northern coast of the Black Sea in Crimea, Ukraine. Both settings offer remarkably well-preserved ancient rural landscapes, once densely occupied by farmers and still containing abundant evidence of their homes, burial grounds, and places of worship. Because their locations were strategically important, they also contain archaeological remains from the Roman through Byzantine eras. Through its research, ICA contributes to a fuller understanding of the settlement, economy, and culture of this previously neglected rural dimension of early western civilization. Metaponto and Chersonesos provide a unique chance to compare rural *chorai* at opposite ends of the colonial Greek world, as well as a valuable opportunity to train students, foster exchanges, and generate international collaboration and good will.

Adjunct organizations

The Center for the Study of Ancient Territories (U.S.A.), *Pidtrymka Chersonesu* (Ukraine), and the *Centro di Agroarcheologia Pantanello* (Italy) are nonprofit organizations founded to support the mission and goals of ICA. Their special focus is on expanding international cooperation for ICA's projects and ensuring cultural awareness and compliance with local laws.

Plate 1. Photo-composite of a Byzantine mosaic on the grounds of the Preserve museum. (See "Documenting the Mosaic Floor of a Byzantine Church," p. 36)

Plate 2. Aerial view of the ancient city of Chersonesos (left), St. Volodymyr's Church (center), and the Museum grounds (right). In the background, the city of Sevastopol.

2002 IN RETROSPECT

Joseph Coleman Carter
Director, Institute of Classical Archaeology

Last year's annual report ran to a hundred pages. This year's might have surpassed that, had we not decided that a smaller volume could convey the major achievements and present the range of our activities, while conserving time that could be better spent writing definitive, rather than preliminary, studies.

Our past year might best be described as "extremely busy." At Chersonesos, Ukraine, and Metaponto, Italy, so many different yet related projects were going on that it was a challenge to keep them all on track and progressing. We had our share of problems—as could be expected when so many people are involved in complex interactions—in an evolving Ukraine and in an ever-changing Italy. The end result, however, was a shared sense of satisfaction coming from successful collaboration. This sense of fellowship was a help to us in the face of the tragedy, acknowledged in the dedication, which affected all members of our group, leaving us deeply saddened but feeling more than ever before like a family.

In recent years, significant grants from the Packard Humanities Institute (PHI) have allowed the level and range of ICA's activities to expand in productive and promising directions. Archaeological research, excavation, survey, and related studies aimed at learning about ancient life—particularly in the Greek and Roman countryside—have remained central to the mission of the Institute of Classical Archaeology (ICA). But we now do a great deal more, a fact that is reflected by the composition of the ICA staff, its collaborators, and chief consultants. Besides archaeologists, paleobotanists and archaeozoologists, there are, for example, architects planning the use of spaces, interior and exterior, a former journalist, responsible for a special publication project, specialists in remote sensing from space and on the ground, conservation experts, budding diplomats, and students of cultural resource management.

The first quarter century of ICA's existence was devoted to making our archaeological work as multi-disciplinary as possible, exploring a multitude of clues, from historic aerial photographs of the chora to the pollen grains preserved in an ancient sanctuary. Now, at the beginning of the new century, the scope has broadened. It now includes, along with the study of past life, the future of archaeological sites and the responsibilities of the communities that have the good fortune to occupy those sites.

Chersonesos

The rapid rate of change in Ukraine has resulted in our devoting more time and effort there than in Italy to the public side of archaeology. We have gained much from the contributions of two high-level consultants, Neal Ascherson, of the University of London, and Dr. Henry Cleere, previously of the International Council on Monuments and Sites, who advised the National Preserve of Tauric Chersonesos on UNESCO's requirements and the steps needed to meet them (Fig. 1). Chersonesos is, in many ways, a model for the future development of archaeological sites in Ukraine and Eastern Europe. We feel that we and our Ukrainian colleagues are writing the textbook for this development.

It is no exaggeration to say that ICA has been involved in most aspects of operation and research at the National Preserve. The challenges there, economic and political as well as scientific, have at times required our urgent attention. ICA, with the support of the Packard Humanities Institute, played a crucial role

Figure 1. Professor Henry Cleere, with the help of interpreter Svetlana Telenkova, addresses staff and students in front of the Chersonesos Museum, explaining the requirements and advantages of a UNESCO World Heritage Site designation. Chersonesos is currently on the short list of candidates in Ukraine for this designation.

in saving the archaeological site—arguably the most important on the Black Sea—from annexation by the Orthodox Church of the Moscow patriarchy. The political solution ultimately adopted by the Ukrainian government was wise and conciliatory, protecting the integrity of the archaeological site in perpetuity as a national preserve and property of the State, while permitting special religious celebrations in a sanctuary restored to its former glory in 2001 (Plate I). This was described at length in the November–December 2002 issue of *Archaeology Magazine,* in an article titled "Legacies of a Slavic Pompeii."

The long-standing economic crisis in Ukraine, exacerbated by rapid privatization of the country's resources and means of production (and centralization in the hands of a few), has resulted in inadequate funding for public entities like the Preserve. This has affected basic maintenance and security, not to mention the desperately-needed conservation of the structures in the chora and ancient city. The government's modest budget for research largely disappeared a dozen years ago and has not reappeared. Yet, in spite of the financial straits, a great deal of research is being carried out by the joint efforts of local and foreign teams.

Further, the major challenges of preserving and conserving ancient structures, both in the city and its unique chora, are being met with immediate action and long-term planning. In 2002, conservation of part of the city wall was carried out, with unexpected but welcome government funding, under the supervision of the Preserve. Conservation experiments were conducted in the southern residential quarter with support from U.S. based non-profit organizations, and plans were made for an extensive intervention throughout the ancient city in 2003 (Fig. 2).

Figure 2. Larissa Sedikova, Deputy Director of the Preserve in charge of conservation (and co-director of the 2002 City excavation), discusses conservation plans with Aleksey Tokarev of Krymrestauratsia. The plans were partially implemented in 2002.

Without the dedication of the Preserve staff and its collaborators, and the generosity of individuals and of foundations in the United States, especially the Packard Humanities Institute, it is certain that Chersonesos' candidacy for UNESCO World Cultural Heritage status, now under serious consideration, would have remained little more than wishful thinking.

In summer 2002, an international team of archaeologists worked to uncover a portion of the residential quarter of Chersonesos, abandoned in the 13th century AD. The goal is to reveal all aspects of everyday life in a city block that was continuously occupied for at least 1500 years. All the techniques of modern archaeology and related disciplines are being brought to bear, and new approaches to recording are being developed on the site.

Our team in 2002 was as multidisciplinary as it was in any of the years at Metaponto. There was fruitful collaboration with two young specialists from Simferopol studying the physical anthropological evidence of the burial population—some 64 individuals from one of the three tombs in and near a chapel in the residential quarter. An important beginning was made in the study of the other organic remains, the ancient plant materials and animal bones. Soil samples were taken within the area of excavation in the city and in the chora as well—at Bezymyannaya and at the Greek and Roman farm, Site 132, where excavation and conservation have been going on simultaneously since 2001. (Both of these areas of the chora are to be nominated for Heritage status.)

The first paleobotanical studies in 15 years to deal with macroscopic remains of plants in the city and chora were led by Dr. Galina Pashkevitch, Ukraine's leading paleobotanist, from the Institute of Archaeology, Kyiv (Fig. 3). The complementary study of animal remains from the chora and those of the city by Dr.

Figure 3. Dr. Galina Pashkevitch uses water-separation to recover plant remains from the ancient city and chora.

Oleg Zhuravlev holds great promise. A part of the faunal remains from Bezymyannaya were examined by Dr. Aleksey Kasparov of the Institute for the Study of Material Culture in St. Petersburg. Both paleobotanical and faunal investigations will continue in the future with a growing number of collaborators. These disciplines are traditional areas of strength in Russian and Ukrainian archaeology, and it is gratifying to see them being revived now at Chersonesos after a long hiatus. The results of these studies will appear in the next annual report.

The picture of the chora of Chersonesos emerging as a result of the paleoenvironmental and geomorphic studies of Paul Lehman and Carlos Cordova, and the remote sensing work of Jessica Trelogan, is making possible a thorough comparison of Chersonesos and Metaponto, the two best and most thoroughly preserved ancient Greek chorai.

The excavation campaign at Bezymyannaya, our fourth year at the site, was undertaken primarily to test the results of an extensive remote sensing campaign in 1999–2001. This year's campaign achieved all of its goals and more, under the supervision of Adam Rabinowitz (University of Michigan) for ICA, and Tatiana Yashaeva for the Preserve. The fieldwork at Bezymyannaya and Site 132 is part of the major joint study of the chora established in 1994 by Dr. Galina Nikolaenko, Deputy Director of the Preserve, and Prof. Joseph Carter, Director of ICA, as general directors and coordinators.

In the city, the international team was co-directed by Dr. Larissa Sedikova, who has directed her own excavation in this area for 15 years, and by Prof. Paul Arthur, an accomplished medieval archaeologist at the University of Lecce in Southern Italy. The student excavators included both veterans and beginners from the Kyiv Mohyla Academy University, Taras Shevchenko University, the University of Simferopol, the University of Lecce, and the University of Texas. Six of the twelve students, like their predecessors in 2001, received credits in the archaeological practicum of Kyiv Mohyla and Institute of Archaeology of the Academy of Sciences in Kyiv. Some students from the United States chose to receive academic credit from the University of Texas for the summer field methods course that is offered annually. Volunteers from Kharkiv and Sevastopol proved quick learners, under guidance from our competent and personable field directors and the more advanced students. The international team of volunteers and consultants, hailing from nine countries, could hardly have been more varied, and I can say without exaggeration it was one of the best humored and most cooperative groups with whom I have ever worked.

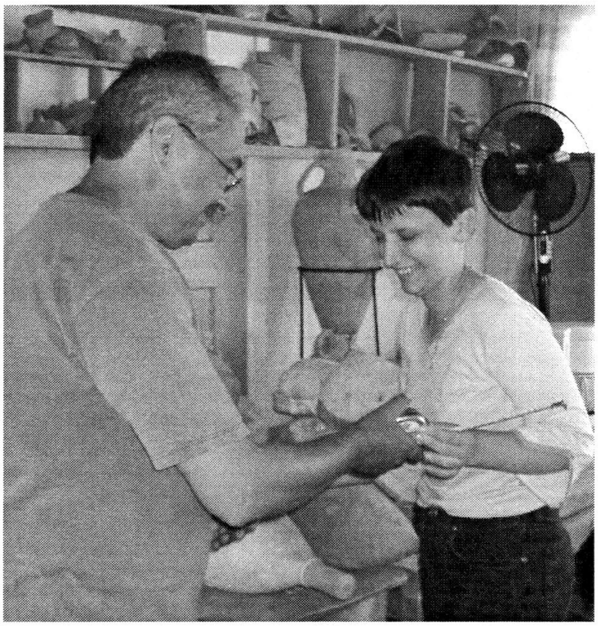

Figure 4. Prof. Andrei Opait, assisted by Anca Dan, conducting a study of Chersonesan amphorae.

The two excavations and several of the research projects carried out during the summer are described in this publication. The major efforts in conservation of the site, preservation of records in the archives, and creation of the Chersonesos web site have received only brief mention. The imbalance will be addressed in the next report, as most of our activity in 2003 is being directed to these areas and to preparation for publication of sites and materials previously excavated. The web site project and the creation of a database for the Preserve's irreplaceable archive (containing records of excavations and research going back to the early 19th century) were the conception of Lucy Grinenko, head librarian of the Preserve Museum; Oleg Vertolitsky, computer system administrator; and Glenn Mack of ICA. Oleg, who died during the summer in a tragic accident, had the satisfaction of learning in the spring that the Ukrainian Minister of Culture proclaimed his web site design the finest for any cultural entity in Ukraine and the model for future web sites. We trust that the database, known as the Megarika project, now carried forward by Oleg's friends and fellow graduates of the Computer Sciences Program at the University of Sevastopol, will reflect his standards of excellence. Despite the loss of this special young colleague, this valuable work continues.

Lucy Grinenko won a grant from the International Board of Research and Exchanges (IREX), Washington, DC, to study modern computer-based archival methods in the United States in fall 2002. She chose the excellent program at the University of Texas, working with supervision from Prof. David Gracy. She contributed as well to the work in ICA's publication projects during her stay in Austin.

In 2002, progress on publication of the Bezymyannaya site moved forward with the addition to our team of Dr. Andrei Opait, an expert in the amphora and coarseware production of the Black Sea area (Fig. 4). Andrei will push to complete the study of the Bezymyannaya pottery in 2003, with assistance from graduate student Anca Dan, student of classics and archaeology at the Sorbonne, and an experienced team of volunteers. Denis Zhuravlev completed his monographic study of the red-gloss pottery from the site. These studies, together with the contributions of Evelina Kravchenko, have established that Bezymyannaya was occupied from the late Bronze Age to the 11th century AD, before being reoccupied in modern times during the Crimean War.

Work also went forward on the growing project of publishing one of the most important groups of monuments of the Greek Colonial period at Chersonesos, the grave monuments of the Early Hellenistic necropolis. These monuments, many painted, had been re-utilized to fortify the Chersonesos city walls—in particular the Tower of Zeno. The number of monuments keeps growing as Dr. Richard Posamentir, of the University of Bochum and the DAI Athens, identifies more pieces dispersed over the Preserve grounds and storerooms in the forty years since their excavation. He, Chris Williams (ICA), and Adela Sobotkova (Czech Republic) have succeeded in completing the catalogue, which now includes 350 entries.

Most contributors to this project met at Chersonesos in 2002: Posamentir, who will concentrate on the stelai and related pieces; Chris Williams, who will photograph the stelai; Dr. Alla Buyskikh, who will write on the architectural elements of *naiskoi;* Professor Paula Perlman, Classics, University of Texas, who will write on the inscriptions; and Joseph Carter, who began the

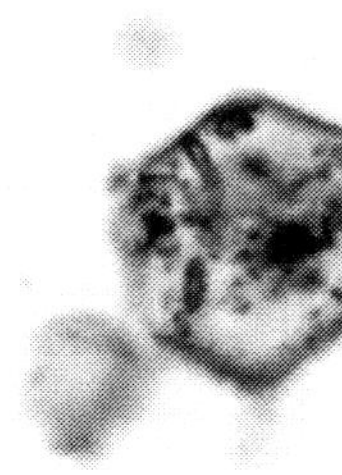

Figure 5. Pigment particle, from John Twilley's study of painted artifacts. (400X magnification)

Figure 6. Architect Carl Holiday's visualization of the interior of the new study and storage facility now being constructed, thanks to a grant from the Packard Humanities Institute.

Figure 7. Monuments from the old storage building were moved to the Greek Museum for safekeeping. The existing Museum display cases were rearranged and the stelai were integrated into the exhibition space.

study in 1994 and will concentrate on the *sarcophagi* and description of the excavation. The group met for several days in early July to coordinate the publication. Another contributor is John Twilley, who is analyzing the pigments used on all classes of monuments at Chersonesos. His results promise to be original and startling (Fig. 5).

A generous grant in 2002 from the Packard Humanities Institute made possible the design and imminent construction of the first building at Chersonesos intended for modern archaeological research and storage (Fig. 6). The site will occupy a completely reconstructed monastery building in the complex that includes the two museums and other buildings of the Preserve. The new laboratory will answer one of the most pressing needs of the Preserve, that of secure storage. The previous storage facility, where the famous painted Early Hellenistic stelai were housed, was on the verge of collapse. In a well-organized, well-planned move, the stelai collection was carried by hand to the Greek Gallery of the Museum, where it will be stored for the duration of construction (Figs. 7, 8, 9)

Last, the proposed development and management plan for the chora, designed by landscape architect Alma Maldonado Holiday and architect Carl Holiday, will soon be printed. It will be followed, in mid-2003, by *Crimean Chersonesos*, the first modern study in English of the whole of Chersonesos, its chora, museums, and environs. It is being assembled and supervised by ICA's Glenn Mack. Both publications are the result of the efforts of a large number of dedicated people.

Metaponto

Metaponto has seemed a second home to me and the generations of students and colleagues that have

Figure 8. The old monastery building, where the funerary monuments were stored, is being extensively remodeled into a storage and research facility.

Figure 9. Temporary metal storage buildings were constructed on the Museum grounds to house the numerous pieces of grave monuments, identified by Richard Posamentir, that had long been separated from the main collection.

come to work and share the experience of living in one of the least known but most fascinating parts of Italy, both ancient and modern. Thanks to a convention between the Region, the Soprintendenza Archeologica—in the person of its dynamic Soprintendente Dr. Maria Luisa Nava—and ICA, a part of several structures of the Azienda Pantanello is on long-term lease as the Centro di Agroarcheologia. Needed restoration took place in 2002 and the newly restored excavation house, office, and lab will be ready for use during the 2003 campaign.

Publication is currently the principal goal of our efforts at Metaponto. The work on the survey of the chora is being coordinated by ICA's Albert Prieto, which is no small job since the contributors are widely scattered. We are fortunate to have in Italy a core group of skilled students of ancient pottery, including Dr. Enzo Cracolici, a leading expert on black gloss pottery of Southern Italy and Metaponto; Cesare Raho, a trained archaeologist with a gift for drawing and photographing the pottery; and Eloisa Vittoria, who with Enzo's tutoring, is becoming an expert in her own right. This reliable and talented group is at work year-around and is joined for periods by Erminia Lapadula, who specializes in Late Roman and Medieval pottery, and by Dr. Anna Lucia Tempesta, whose specialty is architectural terracottas. Two PhD candidates from the University of Sydney, Smadar Gabrieli, who has prepared the cooking ware from the survey and from the Roman tile deposit at Pantanello, and Richard Fletcher, an expert in the application of computer-based programs for the analysis of large ceramic assemblages, are part of a growing Australian contingent that includes long-time collaborators and friends, the physical anthropologists Prof. Maciej and Dr. Renata Henneberg of the University of Adelaide, whose work on the human remains continues. Cesare D'Annibale, of Parks Canada, returned to Pantanello, after an absence of 15 years, to work on the publication of the survey that he began in the summer of 1981. In the meantime, he carried out the ICA-sponsored survey of Croton (1983–1988) and did extensive survey work in Cyprus. Cesare's database of the coarseware from the survey will fill the final gap in our documentation.

The North American contributions to the study team at Pantanello and in the "Banca" include Prof. Rebecca Miller Ammerman, who, together with Mary Malone, is preparing for publication not only the great assemblage of terracotta figurines from the sanctuary at Pantanello, but also those from the survey. Her previous, extensive experience with the terracottas of Poseidonia-Paestum (the sister colony of Metaponto), in addition to her sound scholarship and enthusiasm, make her the ideal person for the job. Mary Malone continues her tireless efforts in cataloging these pieces.

Prof. Mariah Wade's detailed study of pottery—pottery of all types, not only the black gloss but also the important and often-ignored undecorated storage, table, and cooking wares—promises to be worth the time and energy she and her team have put into it. The publication is being carefully designed and, because of Mariah's attention to detail, will be a valuable reference work for all studying and excavating in Southern Italy.

Three elements of the Pantanello Sanctuary volume—the study of the terracottas, the whole corpus of pottery, and the unique discovery of massive evidence for the plant life of the chora during the Greek period by Prof. Lorenzo Costantini and his team at the Institute of Bioarchaeology in Rome—promise to make the book, in its way, as original as the *Necropoleis* volumes.

The amplification of the Jerome Lectures (presented by Joseph Carter in 2000–2001) into a book, *The Discovery of the Ancient Countryside at Metaponto,* has received positive reviewer reports and is scheduled to be published in fall 2004 by the University of Michigan Press. We are particularly happy that the Press has decided to print all of the numerous illustrations in color.

This major effort to bring together all the significant results of a quarter century's exploration of the Metaponto chora in a unified publication series seemed like a dream five years ago. It is exhilarating to see it now coming together. We owe a great debt of gratitude to our excellent collaborators, and to our sponsors in this country, the Brown and Dougherty Foundations and the Packard Humanities Institute, who have made it possible to assemble our extraordinary team. Our hosts, the Soprintendenza Archeologica della Basilicata and the Region of Basilicata, have also contributed in many ways to our efforts and to keeping alive that spirit of shared endeavors that goes back to the first Superintendent of the Region, Prof. Dinu Adamesteanu, who inspired his students and all who came in contact with him. No effort is being spared to insure that ICA's work will continue to reach its ambitious goals and merit the trust and assistance of our generous supporters.

Plate 3a. A portion of the excavators from the City site gather for a photograph on the last day of the season.

Plate 3b. The excavation team at Bezymyannaya in the final week of the dig.

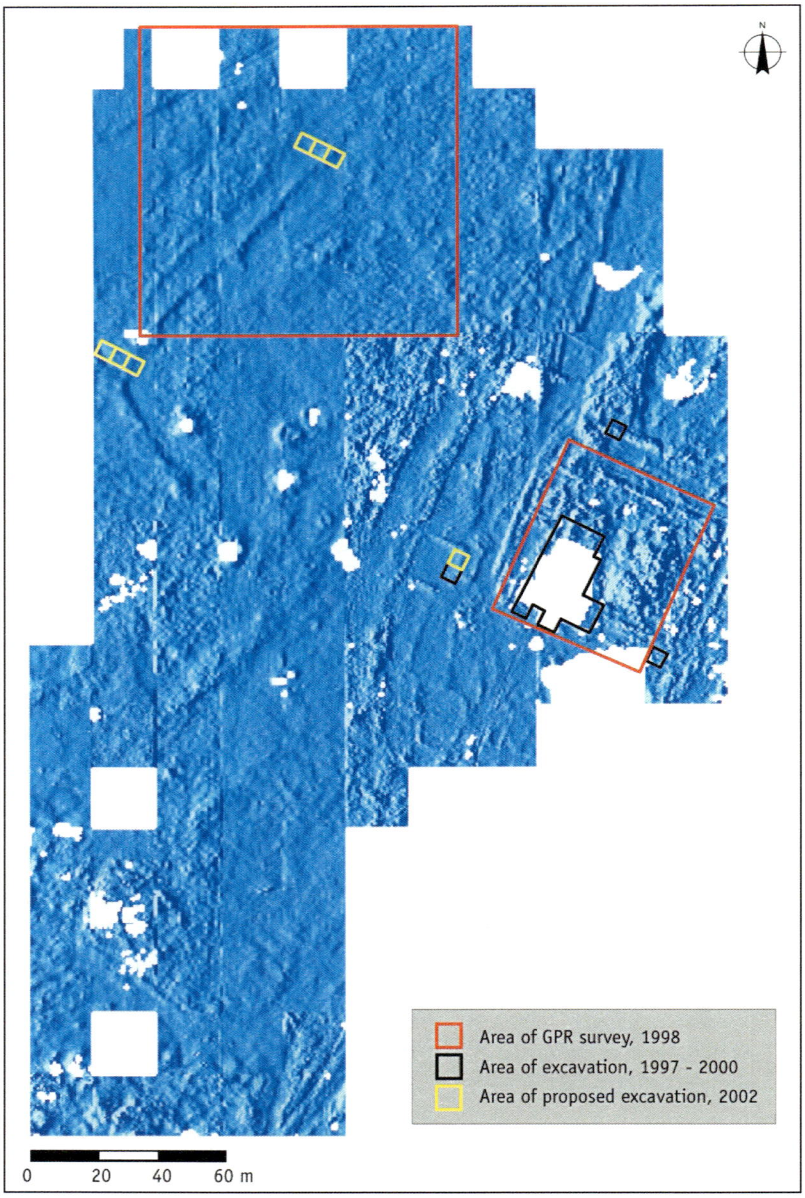

Plate 4. The results of the resistivity survey of Bezymyannaya (1999–2001) carried out by Michail Nikolaenko and Sergi Shakuro. Features invisible on the surface stand out in this three-dimensional rendering. Areas of special interest, explored during the 2002 excavation, are outlined in yellow.

THE CHORA OF CHERSONESOS
EXCAVATIONS AT BEZYMYANNAYA, 2002

Adam Rabinowitz, Tatiana Yashaeva, and Galina Nikolaenko

The site known as "No Name" Hill is visible from the highway as one approaches Sevastopol from the west, commanding the best land pass onto the Herakleian Peninsula, which formed the territory of ancient Chersonesos and modern Sevastopol. It was, as four archaeological campaigns since 1997 have clearly shown, vital to the defenders and to the attackers of these cities. Below it lies the Valley of Balaklava of Crimean War fame. Further to the south, the Taurian Mountains define the scenically spectacular south Crimean coast, home to the indigenous barbarians—the Taurians—when Greek colonists of Chersonesos arrived.

Work began here in 1997 in a short trial excavation by a joint team from the National Preserve, under the general direction of Galina Nikolaenko (the Herakleian Expedition) and Joseph Coleman Carter (the Institute of Classical Archaeology). Longer campaigns in 1998 and 2000 exposed elements of what proved to be defensive positions used in the Hellenistic and Roman periods and well into the Byzantine period. After a hiatus of some 800 years it was overlain and partially reutilized in the mid-19th century by a Crimean War fort, which became the first authentic monument from that period to have been officially excavated. The site was again manned during WWII, during which time it was heavily shelled and damaged.

Four years of geophysical prospection have revealed many deeply buried structures stretching for acres over the northern slope of "No Name" hill. The site is many times larger than it appeared on the surface, though how much larger is still not clear. What purposes and in which periods did these hidden features serve? What can they tell us about the site itself and the chora in general?

Excavation resumed at Bezymyannaya in 2002 for four weeks from mid-June to mid-July. Though time was short, the season was productive, and we managed to achieve a number of goals we set for the project. Foremost among these were (1) the ground-truthing of the extensive remote-sensing (resistivity and ground-penetrating radar) surveys begun in 1998 and completed in 2001 and (2) the further investigation of a possible necropolis uncovered during excavations in 2000. The joint team—led by Adam Rabinowitz and Dr. Tatina Yashaeva and assisted by Gordie Lobay—consisted of a Ukrainian architect, two American students, a Canadian volunteer, eight university students from Kyiv, two long-time Russian volunteers from St. Petersburg, and eight local workmen. In addition, our excavation efforts were complemented by the technical skills of the project's GIS expert, Jessica Trelogan, and her surveying assistant Jamon Vanderhoek, and by Mikhail Nikolaenko and Sergei Shakuro, who were responsible for ongoing resistivity work. Despite its heterogeneous makeup, our team collaborated efficiently, adding to our archaeological knowledge of this rural site. The very name "Bezymyannaya" suggests that the place is unwilling to give up its secrets easily, and our campaign this summer demonstrated yet again the complexity of even its most simple-seeming features.

All members of the expedition were trained students from Ukrainian, western European, and North American universities. The process of teaching the basics of field archaeology and excavation was continuous; by the middle of the season many of the beginners were qualified in carrying out a variety of archaeological operations.

The work and training continued after the field day was over. The finds were processed in the laboratory as the students washed, sorted, and mended fragmentary vessels. Jessica Trelogan and Mikhail Nikolaenko processed instrument data, Adam Rabinowitz and Gordie Lobay compiled a database of excavated materials, and Tatiana Yashaeva worked with documents in the archives and in the library of the reserve in search of analogies in previous work on field systems. Galina Nikolaenko and Joseph Carter, coordinators and heads of this complex expedition, provided general supervision and shared their experience of many years of multidisciplinary work in the chora.

Dr. Andrei Opait launched his major study of the amphorae and other coarseware from Bezymyannaya, while Denis Zhuravlev continued his research on red-gloss ceramics (which is now complete). Dr. Galina Pashkevitch and Dr. Aleksey Kasparov renewed the study of plant and animal remains in the chora; a synthesis of their results will appear in the 2003 ICA annual report.

The beginning of the season presented a number of issues that had to be addressed before full-scale excavation could take place. The first of these was the imminent arrival of a television production crew from Granada TV. The crew intended to produce a segment on the battlefield archaeology of the Crimean War, and it brought several consultants to examine the archaeological evidence for the battle memorialized in Tennyson's "The Charge of the Light Brigade."

Although most of the reports on the excavation at Bezymyannaya have focused on the classical and late antique history of the site, the Crimean War phase is increasingly asserting itself. The hypothesis that the substantial "bank and ditch" fortifying of the hilltop are of the 19th century received support from an English map of Bezymyannaya noting the presence of a "Turkish fort." This was further confirmed by one of Totleben's Crimean War battle maps, showing a square fort with access to the west, located on the hilltop. The excavations of 1998 and 2000 also produced material evidence, including English lead bullets, regimental badges, and a Turkish coin of Sultan Mahmud II of the Ottoman Empire, minted in Istanbul in 1834 it had been pierced and probably used as an amulet by a Turkish soldier. Since the Granada production team expressed interest in Bezymyannaya, we decided to use the opportunity to make a public presentation of the *longue durée* and later history of the site. This decision led early in the season to some clearing of a section of the bank and ditch to the east of areas excavated in 2000; the results are discussed below.

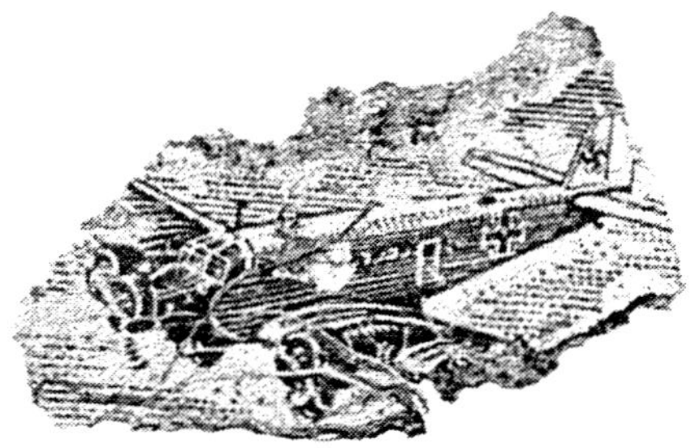

Figure 1. Image from a soldier's papers found in a WWII gun emplacement.

The second major issue was the format of our data-recording system. Though a thorough and sophisticated GIS system of the site had been created and managed by Jessica Trelogan, digital recording of the primary data produced by the excavations had been limited to separate databases for ceramic lots and special finds. It seemed important to reach a certain level of computerization, standardization, and integration this season, before the quantity of data produced became completely unmanageable. The primary categories of data we considered for digitization were stratigraphic information, small finds, and visual documentation (plans and photographs).

Since several different types of information were involved, and since many current excavations are turning towards relational databases (like Microsoft Access), we decided to use this type, in spite of their complexity. To our good fortune, Dr. Giuseppe Gravili, of the University of Lecce, was perfecting his own well-designed Access database for the Department of Medieval Archaeology at Lecce. As a result of his willingness to share with us a copy of his database—modified to fit our needs—we were able to start the season with a data-management system mostly in place. All that was required was a reworking of the paper context sheets employed by the previous site directors. Once these sheets were formatted to reflect the information requested by the database, it was a simple matter to enter the data into the program when we returned from the field at the end of the day. Dr. Gravili's database also allowed us to integrate our small-find data and digital photographs with the basic stratigraphic documentation. Use of this database in the chora will provide a longer-term advantage, in that it corresponds to the database of the urban project, making possible future sharing and comparison of data by the two projects.

We also attempted to more fully integrate our layer plans with the GIS system. In past seasons, graphic documentation of the excavation's progress was more or less limited to top-plans done as sketches or to very rough scale, combined with more thorough final plans produced by the site architects. Utilizing the potential of the GIS to produce accurate plans of significant layers, we attempted detailed 1:50 drawings of each individual layer in addition to sketches and final plans. In some cases, the plans were drawn at 1:50 by American or Ukrainian or Czech students or staff using triangulation, while in other cases the plans were measured drawings done at 1:50 by the site architect. These drawings could be scanned, mapped onto topographic fixed points in the GIS, and overlaid on other digitized geographical information.

The excavation began in several places at once. Two large trenches were begun on the western slope of the hill in order to test remote-sensing results, while a third area was opened just to the north of the trench in which three human depositions were found in 2000. While these trenches were the main focus of excavation, the first days of the campaign also involved the clearing of a 10 m long strip to the east of the area where the body of a WWII soldier was found in 2000. Excavation here was related to the presentation of the Crimean War remains and was limited to the clarification of the structure and composition of the bank and ditch. (This also allowed us to develop a clearer picture of activity in this area during the 1944 German retreat.) It was during this clearing that we uncovered a shallow, triangular cut in the upper layer of the bank, its apex lying just in front of the inside face of the bank. Along that inside face we found parts of a machine-gun loader and dozens of spent cartridges. A more interesting discovery was the leatherette folder with the stamp of the German Reich that almost certainly contained the identification documents of the dead soldier found in 2000. Once restored, the documents may even give him a name. Nearby was a thick card of paper. Like the document folder, it was heavily damaged and almost completely illegible, but we were

Figure 2. A WWII German machine gun emplacement built into a wall from the Crimean War. (Photo: J. Carter)

able to make out the blue-ink print of a WWII German military airplane (Fig. 1). Possibly the soldier was manning a machine-gun emplacement cut into the reused Crimean War bank when he was hit by the jagged chunk of shrapnel that killed him (Fig. 2).

The Crimean War remains also became clearer, and when the film crew arrived, we were able to show them how, in standard fashion, the bank had been constructed with material excavated from the ditch. In the section wall, an uppermost layer of yellow, sandy soil was visible. This soil corresponded with the geological substrate visible in the bottom of the ditch, and was clearly the last bit to be thrown on the breastwork by the 19th century entrenchers. We were also able to exhibit the rough wall visible along the interior face of the bank. This wall had been the subject of some discussion: early hypotheses saw it either as a Byzantine or Medieval wall absorbed by the bank, or as a

Figure 3. A section through a Crimean War rampart and ditch. (Photo: G. Nikolaenko)

Crimean-War period revetment for the bank's inside face. The latter technique was used for some of the less-hasty Crimean War fortifications, and the clearing done during this season suggests even more strongly that the wall, as well as the ditch and bank, belonged to the Crimean War phase (Fig. 3).

Different objects dating to the Crimean War period were found in the context: shrapnel, case-shots, and even a French champagne bottle. It is interesting to note that this section of the bank was built over the Late Byzantine wall. A fragment of a glazed bowl (second half of the 18th to the first half of the 19th century) was discovered when cleaning the wall. This was the first instance in the course of four years of excavation that such late material (from the period soon after the founding of Sevastopol in 1783) has been found at Bezymyannaya's summit.

Andrei Opait's pottery study has documented material from the 11th century AD, relating closely to that found in the city. These important finds extend the ancient, medieval, and modern occupations of the hilltop from the Hellenistic (at least) to the late Byzantine period to our own era. Its life parallels chronologically much of the history of Chersonesos and Sevastopol. The English lead rifle bullet found at the bottom of the inside face of the bank provided a final piece of material evidence for the events examined by the television production team.

Our use of computer applications also interested the Grenada team, as they wished to highlight the technological side of modern archaeology in their program. Jessica Trelogan's development of GIS to overlay, integrate, and study the various phases of the site and the chora, was of particular interest to them.

The other trenches produced material more familiar to the classical archaeologist. One of the two trenches (Trench 1) intended to test the remote sensing results was placed at the apparent intersection between a two-walled linear feature running roughly NE–SW and a substantial division wall perpendicular to it. Here, a ten meter square area composed of four 5 m x 5 m squares was laid out. Time constraints prevented us from excavating the NE square, but the other three were successfully excavated down to bedrock by the end of the season. They revealed a fairly complex stratigraphy composed, in its upper zone, of layers of colluvium from the slope of the hill above, alternating with stony layers representing the gradual collapse of the field walls.

The first two layers—turf layer (context 5201), humus and colluvial layers (context 5202)—were detected at all of the squares. Context 5202 was of particular interest as it contained not only small quarry stones and detritus, but also fragments of Hellenistic ceramics, metallic objects, numerous flint and chert objects—scrapers, knives, cores—fragments of local and handmade ceramics, a fragment of a unique handmade terracotta—an image of a female deity of fertility—and other artifacts which had been gradually washed down the slope as a result of natural erosion processes. Despite this archaeological material not being *in situ*, it is of interest since it reflects a more complex archaeological situation in this part of the Chersonesan chora than has so far been revealed through excavation of undisturbed contexts.

The context immediately beneath these produced a bronze coin *(dichalkov)* dating to the 3rd or 2nd century BC. It was struck in the Chersonesan mint and bears an image of a wreathed head belonging to the Maiden (Parthenos), the main Chersonesan goddess on the obverse side, and the image of a butting bull on the reverse side.

A ten meter length of Wall 1, revealed by the resistivity survey, proved to be 1.0 m wide, preserved to a height of 0.3 to 0.5 meters. The wall's masonry was uncoursed, badly deformed at places, and made of small and medium quarry stones. It can be dated, thanks to the coin and other materials, to the Early Hellenistic period. Below this zone was the NW–SW division wall itself, a substantial but extremely rough construction consisting of a core of small stones and earth with a facing on the south side of larger irregular field stones (Fig. 4).

Figure 4. A length of Wall 1 at the foot of the slope of Bezymyannaya. (Photo: A. Rabinowitz)

Archaeologists working in the Chersonesan chora at different times have had different opinions on the function of such quarry stone walls. For example, eighty years ago L. A. Moiseev saw them largely as land-reclamation structures, while a great expert on the chora, S. F. Strzheletsky, active in the mid-twentieth century, recognized them as a part of a vast project involving land reclamation measures taken by ancient agricultural technicians. Arrangements of roads flanking vineyards, the creation of artificial soil layers, the terracing of steep slopes, the setting of division lines, the erection of protective wind-breaks, and "planting walls" for precipitation collection are all elements of one great coordinating effort to maximize the agricultural production of the chora. A detailed analysis of a number of similar sites has been undertaken by the leading contemporary researcher in ancient Chersonesan chora, Galina Nikolaenko. The debate on "the mysterious constructions of the Herakleian peninsula" recommenced on Bezymyannaya summit in 2002. The function of the wall excavated at Plot 1, however, can be answered definitively only after its complete excavation and an examination of the site as a whole.

Like other walls known from the fields of the chora of Chersonesos, Wall 1 seems to have been built partly as a division wall, partly as a low terrace meant to control drainage and erosion patterns, and partly as a field-clearing wall, such as seen in New England. The wall that this NW–SE wall was supposed to intersect was absent, for the most part, although a thick spread of small stones resting on bedrock probably represents the uphill side of a similar semi-terrace wall. Such irregular packing of stones may have served to drain the walls and prevent them from eroding too quickly. The absence of more substantial remains could be explained by what seem to be heavy erosion patterns in the area and by the immediate proximity of a large bomb crater, the extent of which is clearly visible in the resistivity survey.

Figure 5. Above left, a rare female cult figurine, and, center, an impasto vessel (of the Kizil Koba culture) from the 2002 excavation. The figurine, above right, was found at the site in 1997.

The material from the trench at this intersection provided a great deal of information about the nature and period of use of the field system. The layer of yellowish brown clay (5211), most likely a layer of collapsed mud brick, was 0.2 to 0.7 m deep. It contained fragments of an amphora from Sinope, tableware jugs, and cooking pots of the Late Hellenistic period. Near the foot of the wall, on the bedrock, lay a coin, a *dichalkov,* of Chersonesos minted in 200–190 BC. A fragment of a 4th or early 3rd century BC Attic fish-plate, found imbedded in the division wall, gives a good *terminus post quem* for its construction. On the other end of the chronological scale, the discovery, under the collapse of the wall, of joinable fragments of a Roman amphora of the 2nd or 3rd century AD seems to indicate that the agricultural use of the field system had slowed or stopped by that time.

Fragments of a metal sheath and harness ring found just over bedrock in the stone packing of the main NE–SW wall confirm that the hillside was subject to substantial human manipulation during the construction of the original field system. Moreover, fragments of worn Hellenistic pottery and less worn Roman pottery of the 1st and 2nd centuries AD, collected from all levels of the clay-rich agricultural soil, may indicate relatively deep plowing, at least during the later life of the site. In this area we also uncovered a much earlier object, out of context but nonetheless of great interest: the torso of a handmade terracotta female statuette, similar to a torso found on the top of the hill in 1997, and of a type that seems to belong to indigenous contexts of the Early Iron Age (Fig. 5).

Not all experts, however, agree on the early date of the figurine. According to A. V. Shevchenko, the figurine is a small image of a nude female deity. The position of the hands suggests they might have held certain objects of a cult and, thus, is probably an image of a local goddess of fertility, patroness of people and animals. Despite its primitive appearance, however, Shevchenko dates it to the late 2nd–3rd century AD—a period when coroplasts of the northern Black Sea coast rejected moulds and returned to modeling by hand. Although there were previously no such images in Chersonesos, they are known elsewhere. Such images are typical for barbarian settlements of, for example, Sarmatian ethnic groups, who moved into the chora in the late Hellenistic period.

The whole body of the handmade objects from the hilltop, including the two figurines (this year's and one discovered in 1997) have been studied by E. Kravchenko, who concludes that all were produced by a culture that existed in western Crimea long before the Greeks arrived. If this is correct, the date for the earliest evidence of human occupation at Bezymyannaya can be pushed back into the first half of the first millennium BC.

In addition to the fragments of vessels, a spindle whorl, a round flat artifact with a hole bored through the middle, whetstones, and flint-tools were found in this year's Bezymyannya excavations. With the exception of a few rare fragments of Late Bronze Age ceramics—found in the intermixed levels of the foot of the heights—all the handmade terracotta items from three areas of 2002 field season are, according to Kravchenko, of the same type that characterize the Kizil Koba culture (9th–4th century BC), with which the native population the Greeks called "Taurians" is generally associated.

The second "ground-truthing" trench was located in an area where the resistivity and ground-penetrating radar surveys overlapped (Fig. 6). Our intention was to excavate a cross-section of a linear feature delimited by a pair of parallel walls. This feature was visible on both remote sensing plans, although in somewhat different forms. The resistivity clearly showed two parallel walls, but gave no clue as to depth; the GPR results identified only the denser wall further up the hill, but also indicated that the bulk of this structure was located at a depth of ca. 0.6 to 1 m. Our working hypothesis was that the walls delimited a road—perhaps, from its width (ca. 10 m), an important one. The material discovered in this trench consists of fragments of amphorae and pitchers dating from the 2nd century BC to the 1st century AD. Nikolaenko believes that one context (made up of small yellow stones) could possibly be the remains of an old roadbed. (Its width, 4 m, is a standard for most of the roads at the chora.)

Figure 6. This "ground truthing" trench, number 2, revealed a paved surface, perhaps a road or a livestock passageway. (Photo: A. Sobotkova)

Excavation was to show both the strengths and the weaknesses of interpretations based entirely on remote sensing.

When work ceased in this trench, we had uncovered a 10 m wide strip of bare, irregular, and sometimes jagged bedrock sloping to the west at an angle of ca. ten degrees. On either side of this strip was a rough wall in the form of a pile of medium or large rocks with more or less straight sides. The upper (east) wall clearly served partly as a retaining wall for agricultural soil up the hill; the lower (west) wall was in much poorer condition and seems only to have served to delimit the strip of bare bedrock. Both walls had partially collapsed, and since the collapsed area was difficult to distinguish from the walls themselves, we might not have recognized the features as walls without the information provided by the remote sensing surveys. In this respect, by giving us a larger picture of the area and allowing us to see that these features contained in a regular way—essentially, by telling us what to expect and at what depth to expect it—the remote sensing results were invaluable for the development of an excavation strategy and for the understanding of the area as we excavated.

On an interpretive level, however, results were more mixed. The strip of bedrock we uncovered bore little resemblance to the other main roads known in the chora, which, in general, have well-built stone walls and a smooth and level surface cut down into the bedrock where necessary. The bedrock we uncovered showed no traces of working; in fact, its rough appearance seemed to be the product of the exposure of the natural bedrock, usually smoother, to the effects of the weather. Furthermore, our rock-pile walls were nothing like the faced walls of other roads. At the same time, the strip of land between the two walls did not seem to have been simply delineated as an agricultural zone. The weathered state of the bedrock, the material lying just over it, such as an iron object that may be a link from a chain or harness, and the lack of any soil retained by the lower wall all suggest that the area was bare of soil for its entire working life. What purpose, then, was served by a 10-meter wide zone of bare rock unsuited for wheeled transport? One possibility is that it was intended to be a path for the passage of reasonably large numbers of animals, perhaps to pasture areas located on the inside of the plot. In the end, though, the function of this linear feature remains unknown.

The last trench excavated during this season was located in an area determined both by remote sensing and by the results of previous excavations. On a wide terrace west of, and just below the buildings on the top of the hill, resistivity survey revealed a series of structures (Fig. 7). At the corner of one of these structures, a 5 m x 5 m trench excavated in 2000 had produced the depositions of an adult and two children, and we were excited by the possibility that the area might contain a rural necropolis. To investigate this possibility, we laid out another 5 m x 5 m trench just to the north of the existing excavation. Rather than make the new trench contiguous, however, we took advantage of the resistivity information to locate it slightly further to the north, over the intersection of N–S walls identified in 2000 with the E–W wall of a large square structure to the west. These walls were precisely where the resistivity survey had predicted, but the nature of the area itself was not exactly what we had expected. The layers of soil in the area were all ashy and full of broken pottery, both coarse and fine, fragments of pitchers, amphorae, table- and kitchenware, red-gloss ware, and roof tile as well as large quantities of animal bones showing evidence of butchering. No modern objects were found, which testifies to the purity of the layer. The ceramic material dates the ash layer to the 2nd–3rd centuries AD.

Figure 7. Trench 3 revealed a small structure, possibly used for animals, with a rich deposit of broken pottery.

The only skeleton found during excavation was that of a small dog, apparently left on the surface. There was no evidence of any sort of grave, and although most of the skeleton was articulated, several of the dog's leg bones had been separated from the body, probably during post-mortem predation by other animals. No other bodies were found, and the ashy, artifact-rich deposits covered a beaten-earth surface that seemed to be associated with the N–S wall. This was the earlier of the two walls, and in fact the E–W wall was built on top of some of the lower levels of the ashy deposit

(Fig. 8). In short, as Dr. Steven Thompson suggested after the 2000 excavations, the area seems to have been used as a sheet midden. After cleaning away the stone obstruction, two walls (the northern and the eastern) belonging to different time periods were discovered underneath. The northern wall, whose axis is oriented east–west, was uncovered for 2.5 m. It consists of medium to large quarry stones and its masonry was a double-course with filling. The eastern wall, running from the south–west to the north–east, badly preserved and deformed at places, was exposed for 5 m. It was 0.5–0.8 m wide and similar in construction to the north wall.

The walls both lack foundations, and the size of the structure to which the east wall belongs is such that it is likely to have been an enclosure or pen, rather than a covered building. The smaller rectangular building may have been roofed, but further excavation will be required to clarify the situation. A clue to the use of these two buildings may be found in the results of small-scale flotation performed on samples from the beaten-earth floor: the seeds that could be identified were those of the vetch plant, often used as animal fodder in both classical and modern times.

The material from both the floor and the midden deposit was remarkably uniform in its chronological range. In preliminary readings, the bulk of the material from the lowest levels seemed to belong to the 1st century AD, while the overall impression given by the upper levels was of a 2nd–3rd century AD context. The small amount of residual material present was limited to amphorae and some fineware of the late Hellenistic period. These buildings, therefore, seem to have been built at some time around the turn of the era and soon began to be used for dumping. Neither of them appears to be residential, and it seems reasonable to view these and the other large, irregular structures on this terrace as pens and outbuildings used for purposes of animal husbandry and waste disposal during the Roman period. The absence of a necropolis in this area has been more than compensated for by the vast quantities of interesting material associated with the midden heap.

Figure 8. Aerial view of Bezymyannaya's summit. The upper portion of the photograph shows the southwest corner of the Greek, Roman, and Byzantine fortifications; in the foreground is Trench 3.

The floor of Room 1, of which only the eastern wall was uncovered, had a clay surface. The material discovered on the floor during its final occupation included fragments of red-gloss ceramics, datable to the 2nd century AD. The earlier material found in this layer is remixed, originating from lower layers. The floor makeup was discovered below the level of the eastern wall. It was a dense clay-rich yellowish-red soil which was used to level the natural rock when building Room 1. The layer is 0.05–0.3 m thick. Discovered in the layer was coin of the 3rd–2nd centuries BC, with an image of the wreathed head of the Goddesss Parthenos and a butting bull. The material from the grading layer determined the time of the room's construction. We can distinguish an archaeologically complete red-gloss beaker of the 2nd–3rd centuries AD among the special finds.

Among the pottery recovered in this area is a large sample of fine, red-slipped pottery which may add substantially to our knowledge of Pontic fineware production during the Roman period. In addition, large quantities of handmade pottery were recovered. This pottery has usually been associated with prehistoric indigenous contexts, but its presence in an otherwise unitary and clearly-phased deposit supports remarks made by Prof. Paul Arthur in a previous Annual Report. He suggested that such handmade coarse pottery may represent the persistence of local traditions throughout the Classical period, and study of this material may even allow us to address questions of indigenous–Greek interaction during the later phases of the site. If the handmade pottery is later, then perhaps the two statuettes are later as well, and Shevchenko's date of them is correct.

Figure 9. Worked stone material from the midden in Trench 3.

Related questions may be raised by the small but consistent sample of lithics recovered from the midden (Fig. 9). Unlike the chipped flint tools found in the erosion layers of the trenches on the slope, a large proportion of these lithics seemed to be cores. Given the lack of other residual material in the deposit, as well as the surprisingly high ratio of cores to tools, it seems likely that these flints represent activity contemporary with the formation of the deposit. Such an interpretation is in keeping with the nature of the deposit: one might expect the useless core to be thrown on the midden heap while the tools themselves were used elsewhere. Without anticipating the results of a specialist's work on these objects, we might consider the possibility of an ongoing lithic tradition representing the manufacture of simple tool parts (e.g., the bits of abrasive material used on the bottom of a threshing sledge to separate the grain from its husks) when metal was expensive or unavailable.

Even a productive season of archaeological fieldwork can raise more questions than it answers. The 2002 season at Bezymyannaya is a case in point. Although our sense of the correlation between the structures in the field and the remote sensing results is now clearer, and although we now have a better idea of the chronological range of agricultural activity, only further excavation can provide a better understanding of the patterns of use in the plot. The relation of the midden explored during this season to the three bodies found in the same area in 2000 is equally perplexing, and a combination of forensic examination of the remains and more extensive excavation may lead us to interesting conclusions about marginal burials. A thorough study of each of the categories of material recovered from the midden could also provide invaluable information about the chronology, diet, pottery styles, architectural arrangements, and subsistence strategies of the people who lived on this land in the 1st and 2nd centuries AD. This sort of study and the formal publication of the results are our immediate tasks. When we have gathered what information we can from the material we have already excavated, Bezymyannaya will be as willing as ever to reward new excavations with even more new questions.

From Landscape to Object: Remote Sensing and GIS

Jessica Trelogan
ICA

The results of a NASA-funded pilot project, begun in 1998 in collaboration with the University of Texas Center for Space Research (CSR) and the National Preserve of Tauric Chersonesos, continue to provide extremely valuable information and tools for our increasingly interdisciplinary and international team.[1] In order to investigate the use of remotely sensed data for the study and protection of Chersonesos and its chora, a wide range of multi-temporal and multi-resolution data was obtained from space-based, airborne, and ground-based sensors spanning the last half-century. Where previously there were none available, these data have provided a long-needed mapping base for the large variety of research and management initiatives conducted by our joint project. They have not only vastly improved our ability to map the archaeological landscape, but have also provided a means for the continued monitoring of the site within the region's highly dynamic modern landscape.

A handful of historic aerial photographs from the 1960s, as well as a set of recently declassified CORONA satellite photographs from the 1960s and 1970s, were used to update and improve existing maps of Chersonesos' distinctively divided landscape (see *Mapping the Ancient Territory: Remote Sensing and GIS at the Chora of Chersonesos,* 2001 Annual Report).

More recent data are currently being used to monitor landscape modifications and their effect on the ancient landscape over the last three decades. A series of land use/land cover classifications were created from multispectral Landsat Thematic Mapper (TM) and Enhanced Thematic Mapper (ETM+) satellite data for this purpose. The 30-meter resolution of these data is too coarse to identify archaeological remains (Plate 5a), but their increased spectral resolution allows for relatively rapid automated classification of land cover types, and their temporal resolution (data were obtained for the project from 1984, 1988, 1992, and 2000) allows for monitoring of landscape change using one consistent data type. The mapping products that result from this type of analysis not only provide insight into the dynamics of the modern landscape, but they also provide an extremely powerful visualization of the human and natural hazards to the preservation of the rich cultural heritage at Chersonesos.

The resulting digital thematic classifications can be easily collapsed from specific (such as, for example, a 10-class map, Plate 5b) to general (Plate 6a) levels of detail, depending on the purpose of the map. In the time-series shown in Plate 6a, developed areas that were separated, in Plate 5b, into urban/industrial, and old and new housing (dacha) developments, have been collapsed into one general class, labeled urban/built up, in order to emphasize the combined effects of human activity on the region's archaeological heritage. While data, information, and the written and spoken word are essential to making the case for positive legislation, nothing speaks louder than a large red mass seemingly eating up the landscape. These types of thematic map products have the potential to be highly influential at the local and national governmental levels where decisions about land and resource management are made. The result is a legible and powerful visualization. This series of maps, in fact, when shown to the Mayor of Sevastopol this summer, did produce quite a reaction indeed.

In addition to the lower spatial resolution Landsat data, a panchromatic (black and white) one-meter resolution IKONOS image was acquired for the project with funds generously provided by the Packard Humanities Institute. This image, while by far the most expensive single dataset obtained for the project (approximately $12,000 for a scene that covered the entire peninsula), has proven to be well worth the expense. In fact, even if the military sensitivity of this area did not preclude it entirely, it would take years and many more thousands of dollars to conduct the ground work required to map at the level of detail that this imagery has made possible.

This scene served as the basis for the incredibly detailed land use maps that were created by Alma Maldonado for the planning document, *Development Plan for the Chora* (Plate 6b). This level of mapping is absolutely crucial for the analysis of roads and infrastructure required for the creation of an effective management plan, and it would not have been possible without this imagery. The IKONOS image has also been absolutely invaluable as a field reference map for much of the on-going field work, including excavation and paleoenvironmental research.

Although, throughout the remote sensing project, our efforts have been considerably hampered by our inability to obtain any Global Positioning System (GPS) data in this militarily sensitive area, the use of space-based data to create a consistent spatial framework for

---

[1] Solid Earth and Natural Hazards Program, Grant No. NAG5-7693.

mapping in the chora has vastly improved the previous situation. Unfortunately, without the use of accurate GPS data, only landscape-scale analysis has been possible to integrate into one consistent coordinate system. This has, however, been a significant development for much of the work going on at Chersonesos. It has been absolutely crucial for the study and integration of data relating to the ancient property division system, as well as for the development of a comprehensive management plan for this vulnerable landscape.

Ideally, the landscape level analysis and data man-age-ment could be seamlessly integrated with more detailed spatial data produced by excavation and other archaeological fieldwork (from paleoenvironmental research to building conservation). With the use of just a few precise GPS coordinates, we would be able to fully integrate these different scales of data within a digital GIS. Plans for a GPS survey were underway for the summer 2002 season, but attempts to obtain permission were again unsuccessful. We are continuing to map current excavations with exacting detail (down to individual objects), using traditional total station methods, but with local, arbitrary coordinate systems set up at individual sites (see *Excavation GIS*, 2001 Annual Report). The local surveys are, fortunately, tightly controlled with benchmarks that can be relocated from season to season. As soon as we do obtain permission to use GPS, it will be relatively easy to tie in all the local survey data from previous years of excavation to real-world coordinates. This will allow us to manage all of the relevant current and previous data in one multi-scale GIS, making it possible to "zoom in" from the scale of the landscape, to sites within the landscape, to the objects within the site.

Remote sensing, especially in the form of aerial photography, has had a long history in archaeological research and, to some extent, in cultural heritage management; however, the potential of space-based remotely sensed data has only recently been realized for these applications. This has been, in part, due to the limits of spatial and spectral resolution of the most widely available satellite imagery. These limitations are beginning to become less of an issue due to the recent availability of very high resolution data such as IKONOS, as well as the recent declassification CORONA data.

Very high resolution sensors are indeed more appropriate for the detailed type of mapping associated with small archaeological sites. However, the use of medium resolution satellite data for monitoring archaeological landscapes has also seen an increase in recent years. This is indicative of a number of trends within the fields of archaeology and cultural heritage management, which show a relatively new appreciation for the importance of understanding and preserving cultural landscapes, as opposed to disconnected sites.

Moreover, as computer processing power has increased and desktop based GIS and image processing software becomes more user-friendly and affordable, these technologies are likely to be increasingly tested and improved for archaeological research and heritage management.

While the use of these technologies has been extremely productive within our joint project, our ultimate goal is the widest possible dissemination of information and visualization through web-based publishing. In the meantime, it is one of ICA's main missions to make these types of technologies accessible and useful for Chersonesos' own local stakeholders. This means not only educating the local museum staff and archaeologists in the use of the relevant data and software, but that it is essential that the development of methodologies also be guided by people who will be adopting it on the ground.

The atmosphere of learning at Chersonesos this season was more exciting than ever. All of the student volunteers and a number of museum staff members were eager to receive hands-on training in the use of the total station, and several workshop-style presentations on the use of the database, GIS and remote sensing methods were enthusiastically attended. We are excited to continue this educational outreach in the coming years, and are hopeful that as we continue to develop these methodologies along with our Ukrainian colleagues, the exchange of information and ideas will continue to keep Chersonesos among the world's most treasured cultural sites.

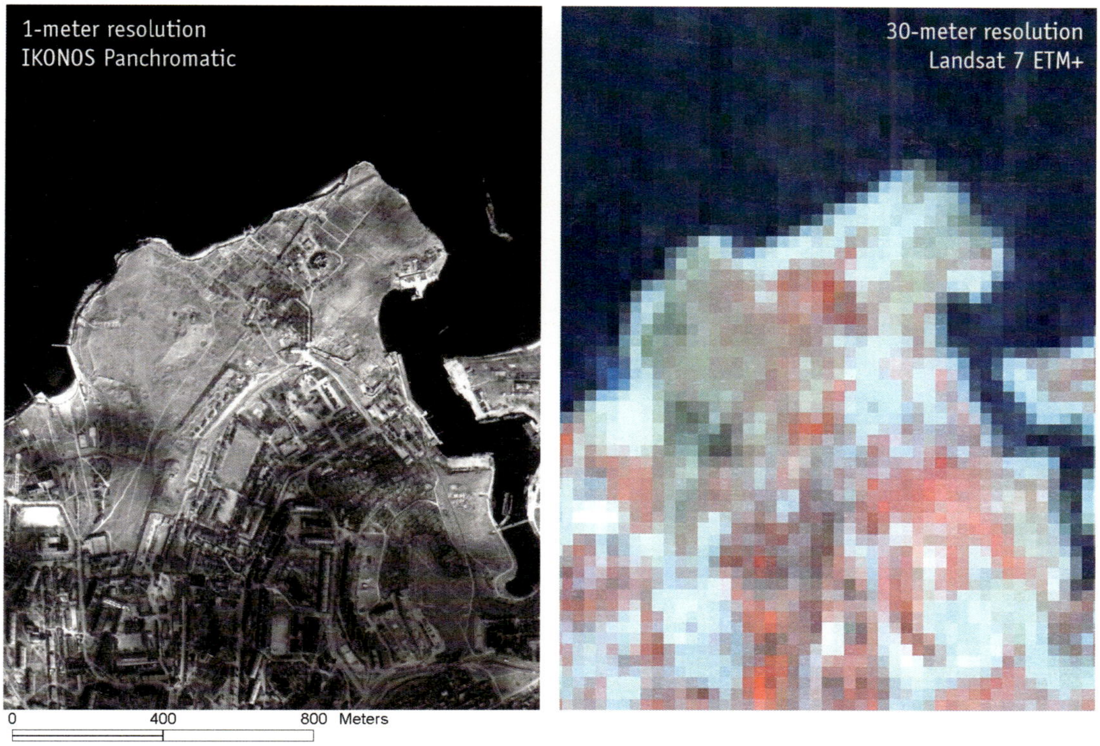

Plate 5a. Comparison of image resolutions: one-meter panchromatic IKONOS (left) vs. 30-meter Landsat 7 ETM+ data.

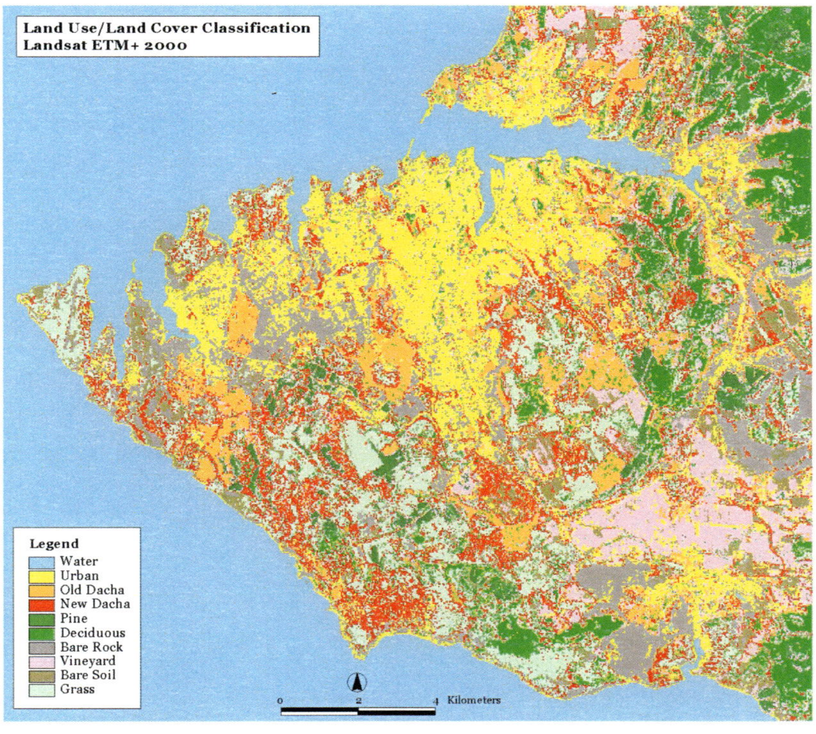

Plate 5b. Detailed (ten-class) thematic land use/land cover map, created from classification of Landsat 7 ETM+ data acquired in July 2000.

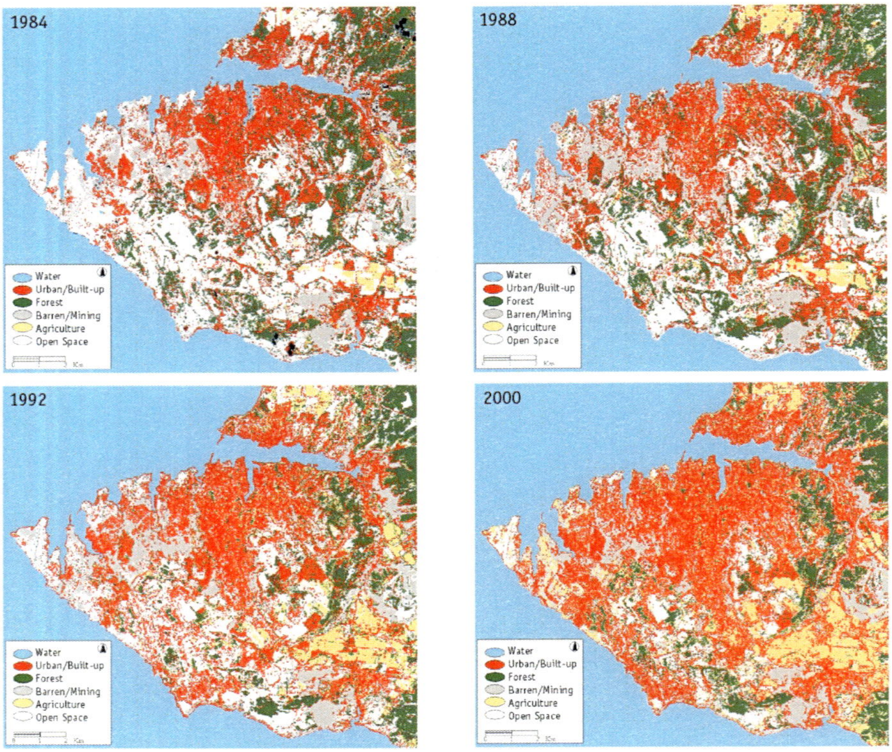

Plate 6a Land use/land cover change detection series. These generalized (six-class) thematic maps, created from classifications of Landsat TM data acquired in 1982, 1984, 1988, and 2000, emphasize the massive urbanization that threatens the preservation of Chersonesos.

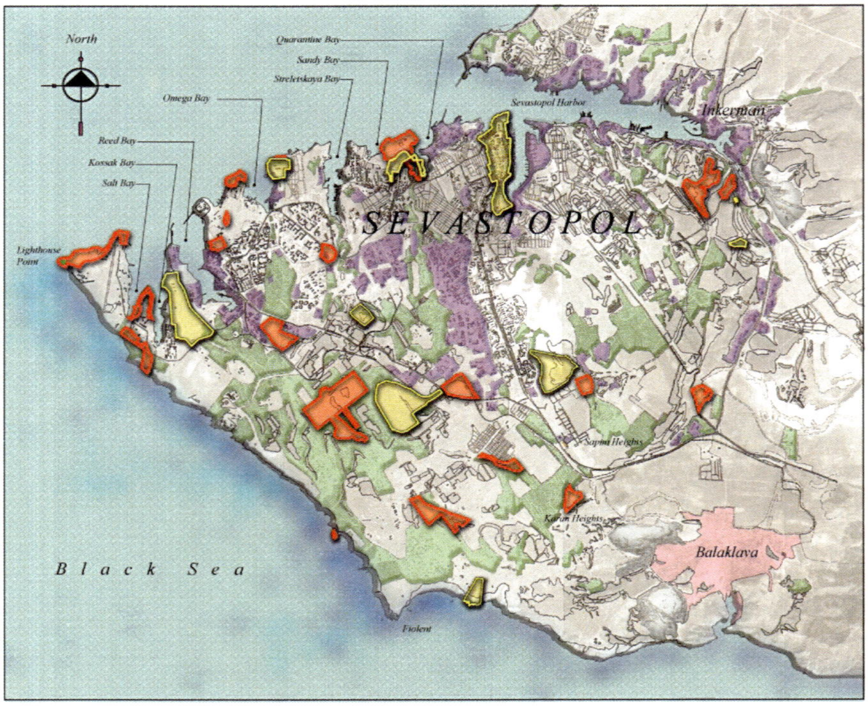

Plate 6b. This illustration, from the *Development Plan for the Chora*, by Alma Maldonado Holiday and Carl Holiday, used an IKONOS image as a base for applying map information.

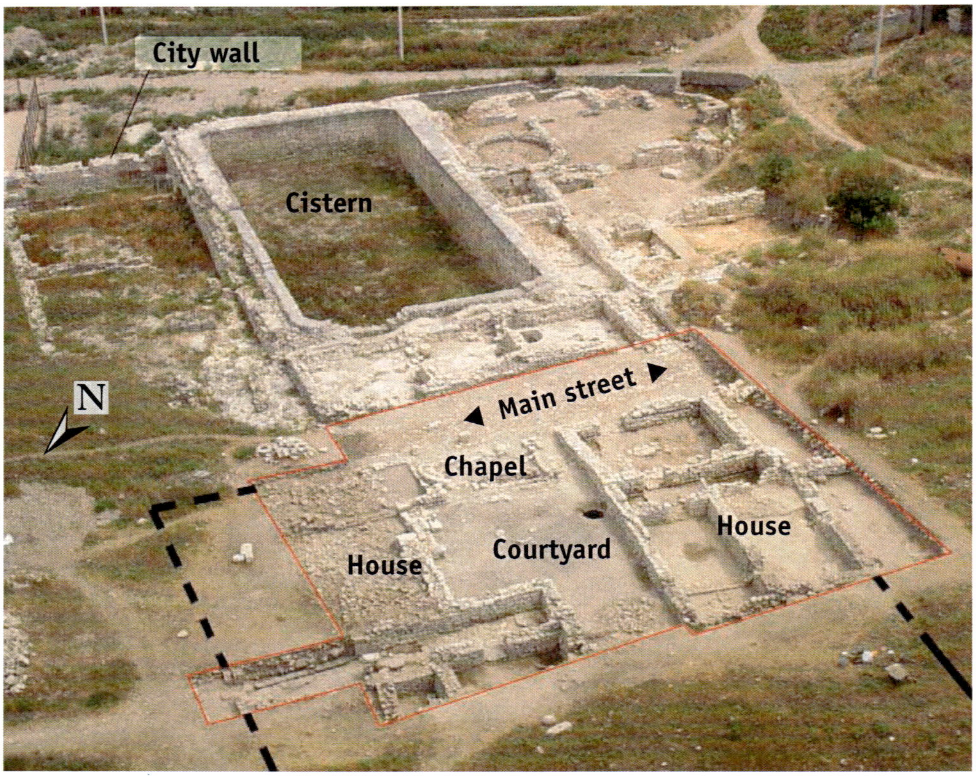

Plate 7. An aerial view of the City excavation at the end of season.

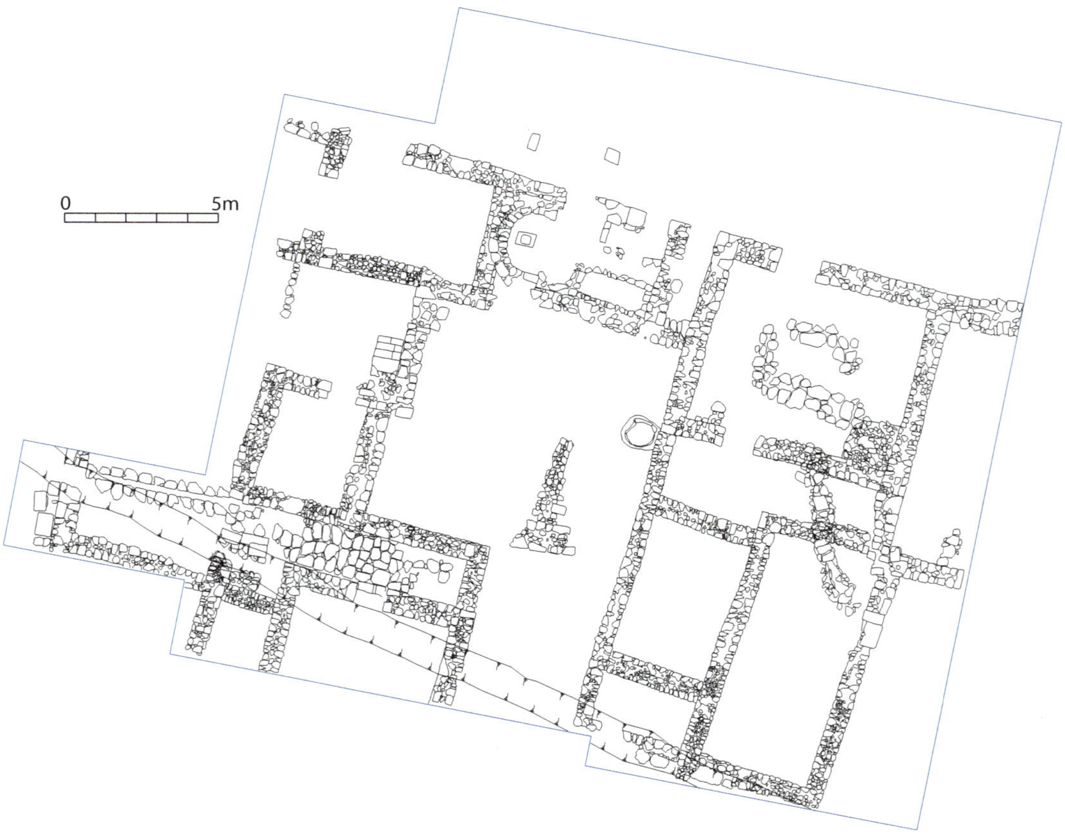

Figure 1. Plan of area excavated in 2002.

EXCAVATION IN THE SOUTHERN RESIDENTIAL SECTION OF CHERSONESOS

P. Arthur, L. Sedikova

The archaeological research of an insula or a block in the southern region of the ancient city of Chersonesos is the continuation of a joint project initiated in 2001 by the Institute of Classical Archaeology of the University of Texas at Austin (ICA) and the National Preserve of Tauric Chersonesos (the Preserve), with the participation of a group from Lecce University (Italy). Medieval Chersonesos, or Cherson as it was called by contemporaries, was a remote but highly significant outpost of the Byzantine Empire on its northeastern frontier. Acting as a link between the Empire and the barbarian world, the city always remained in the thick of things as far as the domestic and foreign policy affairs of the Empire were concerned. However, as an independent city, Chersonesos was a constituent of Byzantine culture, involved in the mysteries and contradictions peculiar to that world. As Chersonesos cannot be fully understood out of its Byzantine context, so the Byzantine Empire should not be limited to the exploration of major centers and their suburbs. The study of a Byzantine province or city in its extreme reaches, such as Chersonesos, makes it possible to glean a more complete idea of the Empire's problems and achievements. The uniqueness of Chersonesos, in comparison to other centers of the Empire, lies in the fact that the territory of the ancient city was practically never built over by later constructions. Thus it remains in much the same state as when it was abandoned by its occupants. This excavation is of particular interest because its object is a residential quarter, not a public building or church. Much is known archaeologically about monumental Byzantine architecture, but almost nothing of private residences and everyday life.

The main objectives of the 2002 excavation season were to define the extension of the Byzantine structures identified in 2001 as lying immediately to the north of a main street and to clarify their chronology and function. Work was thus concentrated on both enlarging the excavation area and of excavating down to earlier layers in the area previously brought to light. The research was successful in defining part of the *insula* block (see Fig. 1), and in recovering further information on the buildings and open areas, as well as in identifying a structure that apparently antedates Byzantine occupation of the area. Together with the many objects that have been found, the evidence appears to show substantial economic growth within the city, particularly as regards building, from the end of the first millennium onwards. These topics are all treated in detail below.

Complex I

What is here termed "Complex 1" is the large building, identified in 2001, that opens directly onto the main east-west plateia, the "Main Street," in the southern part of the city. The complex, composed of five or six ground-floor rooms, appears to have developed around an original free-standing square building. This building, later to become room 26 of the complex, was timber-framed with beams of oak. It is possibly of 10th century date and is certainly later than a bell-shaped silo, which yielded 10th century ceramics, that lies at its north-east corner. The building was gradually added to during the course of the 11th–12th centuries, becoming just one room of a larger complex. Altogether, it measures 36 square meters.

Careful excavation has revealed various details of the room. It possessed a pavement of limestone blocks towards the center, giving the impression that its roof may have possessed a central opening, either a cellar or *compluvium*, as in the *atria* of Roman houses, permitting the entrance of both light and rainwater. The second interpretation is backed by the fact that the pavement has no blocks in its center and has partially subsided, suggesting that a square cistern, which could have been filled by rainwater, may lie beneath. Further excavation is required to resolve this question.

On one side of the paving was a circular hearth, while nearby, in the south-west corner of the room, was a rectangular cooking-platform formed of pressed clay. A central opening in the roof of the room would thus also have provided an outlet for cooking smoke. The north-west corner of the room was occupied by a substantial limestone platform, intended to support a large jar or *pithos*, from which water or some other liquid could have been readily tapped. Excavation of the south-east corner of the room yielded a fair quantity of smashed glazed dishes and bowls, perhaps from collapsed shelves. Numerous fragments of fine ware, including Zeuxippus ware ceramics, were found on the floor in the northeastern part of the room. This group of luxurious glazed ceramics was widespread in Byzantium; however, all of its manufacturing centers have not yet been discovered. White clay glazed vessels with painted decoration were most probably made in the capital. Syrian bowls with turquoise glazed coating were also found in the same room (Plate 8a). A golden ring-shaped earring with a cross-shaped pendant was discovered in the same layer (Plate 8b). The evidence would seem to indicate that room 26 was used primarily for the preparation and serving of food.

On the opposite side of the street entrance, room 26 opened onto a rectangular corridor (room 27), which, among other things, yielded a stone rotary hand-mill for the preparation of flour. This room, in turn, opened onto another (room 21), which appears to have been used as a storeroom. Within it were numerous wine amphorae and storage jars, smashed by the falling roof. This room, which originally was two rooms, was created by the demolition of the dividing wall.

Though no staircase has been identified in complex 1, it is feasible that living quarters were located on a second floor. However, the fact that rooms 26–27, 21–23 were apparently all dedicated to the storage and preparation of food may suggest that the complex, at least in its last phase, functioned as some form of restaurant or *taberna*.

The Yard

Complex 1 was bordered to its immediate east by a yard. This space was separated from the road by a small church and closed to its east by what appears to be two further, distinct buildings (complexes 2 and 3, below). The earth surface of the yard presented extensive patches of burnt material, charcoal, and ashes, as well as an impressive concentration of iron slag fragments, of which some 4 kilos have so far been collected, indicating that iron working took place there. Evidently, some inhabitants of the insula were involved in metalwork; however, we were not able to discover what kind of objects they manufactured. We hope that future excavation of the rooms northeast of the yard, designated complex 3, will help us learn more about this craft. This season only the turf layer was removed and room boundaries were outlined in this part of the block.

The Church and its Burials

Between the courtyard and the *plateia* a small church had been erected, housing three rectangular stone-build tombs. Other burials or charnel pits may lie outside the church. One of the tombs was meticulously excavated in 2002, bringing to light the remains of 64 individuals. Some thirteen of them were articulated, including 8 adult males, while the skulls of many of the disarticulated skeletons had been respectfully heaped in the western end of the tomb. A few objects were discovered in the grave: bronze button-bells, a silver ring-shaped earring, and several bronze Chersonesan coins dating to the 13th century.

Preliminary analysis of the skeletal remains by Denis Ponomaryov and Vladimir Smirnov shows a high infant mortality (44% of the sample), as well as three fetuses, and a quite remarkable range of adult pathologies (see Preliminary report on Physical Anthropology). As it is suspected that the church may have served the community inhabiting the *insula* and its immediate surrounds, it will be interesting to see if there are any common genetic traits among the individuals. In any case, it is clear that a more detailed analysis of the human remains will provide important information on Byzantine population, demography, and burial.

Complex 2

Excavation in 2002 of the northeastern part of the block was complicated by a deep trench discovered along the new site. Its origin can be associated with the construction of a water pipeline dating to the post-World War II period. The trench badly damaged ancient layers, but we still managed to complete the examination of room 25 discovered last year, and to uncover partially rooms 30, 31, and some open space, probably a yard, in the northern part of Complex 3.

Room 25 turned out not to be a cellar as we had earlier supposed, but a long corridor paved with limestone slabs, which ended in a courtyard. The remains of a burnt wooden door and numerous metal fixtures were discovered outside the entrance. Because the floor surfaces of the room and the courtyard are both at the same level, we can suppose that the bedrock on which rests the northeastern part of the assemblage has natural terraces receding toward the center of the ancient city.

Room 25, a source of great excitement last year with the discovery there of remarkable ceramic finds including a unique ceramic cooking stand, justified our continued hopes this season. First of all, we uncovered some missing fragments of the stand, and several polychrome glazed pottery fragments that joined vessels discovered last year. Still, the most impressive find of the season was a small stone icon depicting the bust of a beardless Christ with a halo made of crosses (See front cover). Similar icons were probably worn hanging from the neck over clothes. This is the thirteenth steatite icon to have been found in Chersonesos. Most of these date to the 11th–12th centuries.

Rooms 25 had two more exits—a broad doorway leading to room 30 and a staircase leading to room 31. Room 30 was not completely excavated, but the floor layer was sufficiently distinguished to have a large ash spot and sooty stone blocks, one of them with an etched cross. This was most probably a fireplace. It is interesting to note that there were some grooves, probably drains, leading from room 25 to room 30 and to the courtyard. The abundant remains of fish bones and scales found in room 25 suggest that gutting may have taken place there which, in turn, might explain the presence of drains. Like building complex 3, it had a flight of stairs clearly leading to a second storey. Two-storeyed buildings were apparently com-

Plate 8a. Turquoise glazed bowl reconstructed from pieces found in Room 26. (Photo: G . Mack)

Plate 8b. Gold earring with a cross pendant, from Room 26. Length: 8.5 cm.

Plate 9a. The processional cross found in 2001. Conservation revealed silver inlay. Height: 47.5 cm.

Plate 9b. Two small finds: (left) intricately carved bone, length 4.3 cm; (right) part of a glass cup handle, length, 2.6 cm.

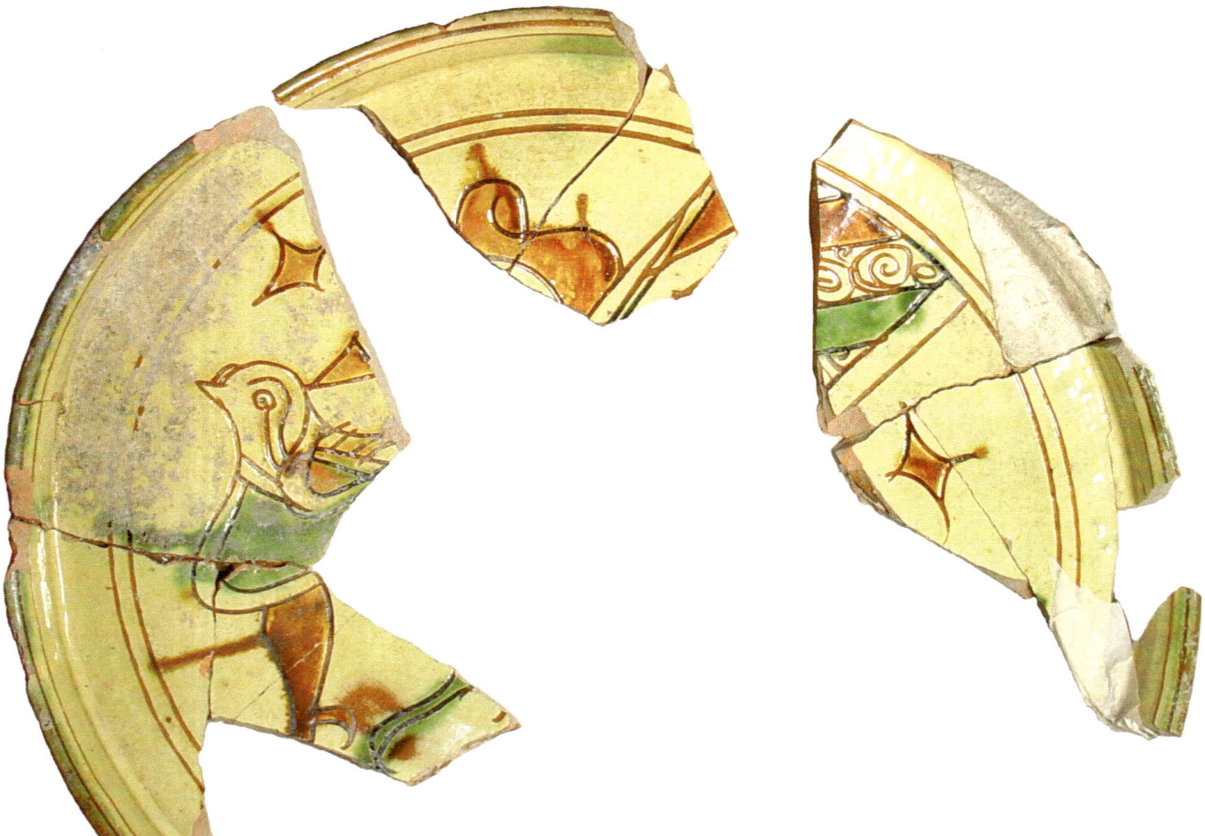

Plate 9c. Graffito ware plate, ca. 13th c. AD. Diameter, approx. 40 cm.

mon at Chersonesos. It may have been that both room 25 and room 30 with its fireplace were in some way involved in the treatment of fish. Tile fragments with relief letter marks and horseman images were discovered in the layers of collapse, indicating the rooms had tile roofs.

Figure 2. Two column capitals, possibly from an early Christian church, found on the site's eastern edge.

Judging by a small strip examined in the northern part of the excavation site, we concluded that the block was separated from the street, the presumed east stenopos, by a massive wall with an entrance. This strip, excavated down to the layer of the last usage of the street, revealed interesting finds. There were two marble slabs probably used as benches at the entrance to the courtyard. When cleaned, they turned out to be impost capitals of columns. They are made of white, probably Proconessian, marble, decorated with carved volutes and acanthus leaves (Fig. 2). Crosses are cut on the impost side of one of the capitals. It is absolutely clear that these unprofessionally cut crosses appeared after the original use of the architectural elements. Undoubtedly, these capitals were originally parts of columns, that decorated the interior of a Christian church. These finds, together with other fragments of marble architectural elements discovered at this site last year, make it possible to suppose that there was an early Christian church, previously unknown in Chersonesos, situated nearby. The majority of similar architectural elements belong to basilicas, which existed in Chersonesos in the 6th–10th centuries. The question now is, will more of it be found as excavations are extended. If so, it would be the first basilica in Chersonesos to have been excavated using modern criteria and methods.

Street Plan

The "Main Street" or *plateia*, running east through the southern part of the city, is over 6 m or about 20 feet wide and was brought to light during the 2001 excavation season (see Street Plan, Fig. 3). As it is clearly aligned on the rectangular street grid that characterizes the layout of Chersonesos and its chora, it was reasonable to assume that secondary streets or *stenopoi* led off it at right angles, thus defining a rectangular *insula* block. Excavations this year, as noted above, appear to have uncovered one of these *stenopoi* at the eastern edge of the site.

Given that the widths of the *insulae* at Chersonesos are generally around 26 meters, the location of a second *stenopos* to the west has been plausibly identified in the area already unearthed as a space later occupied by buildings, rooms 20 and 22, which were cleaned down to the road surface at the beginning of the 2002 campaign. Apart from deductions based on measurements, this second *stenopos* would appear to be aligned

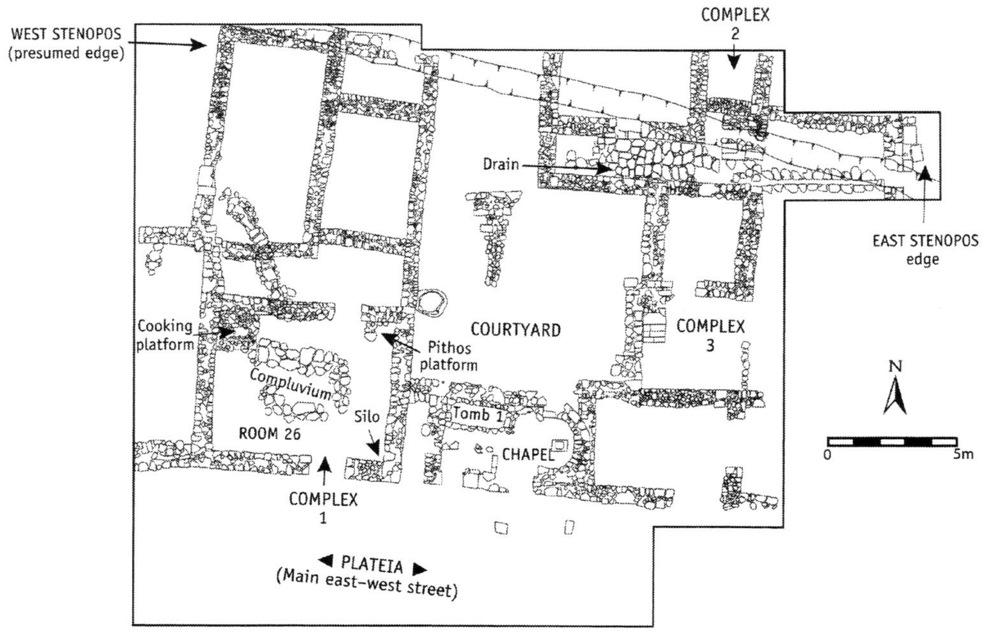

Figure 3. Chersonesos city: insula opposite cistern.

on a paved alley to the south of the Main Street, leading to a building sited to the west of the town's public reservoir, already excavated by Larissa Sedikova in the 1990s. The main street sloped downward from west to east. It has now been ascertained that the difference in levels indicated by the street was mediated within the *insula* by terracing so that the various buildings opened onto the streets at different levels.

Pre-Byzantine Remains

The earliest remains identified in the area of the excavation consist of a short section of dry-stonewall constructed of large, irregular, limestone blocks, which was later truncated by the walls of a Byzantine building (complex 1). Without further excavation, little else can be said of this structure at present, save for the fact that it appears to be associated with ceramics of Roman imperial date. All further remains uncovered date to Middle and Late Byzantine times. One of the goals of the project is to explore the relationship of this area of the late Byzantine city and its predecessors and, in particular, to learn whether the latest street plan follows closely that of the colony—presumably laid out as a grid in the 4th century BC according to the so-called "Hippodamian" systems. This was used on a much grander scale to divide the chora at that time.

Date of Destruction

The whole insula appears to have been destroyed during the second half of the 13th century, but not before the secondary roads or *stenopoi* had been blocked by cross-walls, as in other Byzantine towns. So sudden must have been the destruction that many objects were left in place for us to uncover. Complex 1 yielded a storeroom full of smashed commercial transport amphorae and *pithoi*, as well as an iron processional cross, found in 2001 (see Plate 9a). Conservation has now revealed that it was decorated with silver inlays. Both complexes 1 and 2 have provided some splendid examples of 13th century glazed pottery, including a range of high-quality: the Zeuxippus Ware vases (known from the baths of the same name in Constantinople), polychrome Glazed White Wares (GWW), local sgraffito wares (see Plate 9c) and a range of champléve wares, probably from Thessalonika.

Archaeological Documentation

The absolute wealth of finds now counts over 2,000 items mapped with three coordinates, using a total-station, and includes many coins. The thousands of pottery fragments, animal bones (mainly sheep/goat and fish) and other objects have been positioned according to their stratigraphic context. A specific database and GIS mapping system has been created to handle the data, which includes hundreds of documentary records, site plans, artifact drawings, and digital images. This is the first excavation conducted at Chersonesos making substantial use of modern computer-based technology. This allows great precision and analytical capabilities that were not possible on any excavation even ten years ago.

Conclusions

The 2002 excavation continued the examination of a residential block in the southern region of Chersonesos. Its southern, western, and eastern borders were defined. The part of the insula uncovered so far consists of at least two separate estates divided by a wall. For the time being, complex 3, which is essentially unexcavated, has been assumed to be a separate residence. Only further excavation will clarify this. The eastern estate, if it proves to be a single unit, would have been twice as large as the western one. They were rebuilt more than once and, as finds indicate, were abandoned not earlier than the second half of the 13th century. What historical events caused the demise of this block? Was it a Tatar invasion? We hope to find answers to these and other questions in the course of further research. The excavations to follow will make it possible to complete the analysis of the block and to explore earlier layers and structures. Because the site is located on an arterial road of Chersonesos, we might suppose that excavation of earlier layers will provide finds of no less interest than those of this field season. As regards this year's finds, their detailed study will give us an opportunity not only to imagine the life lived in a block in Chersonesos, but also to get a clearer idea of medieval society as a whole.

Preliminary Report on the Physical Anthropology Studies of the Materials from the Excavations of a Church in the Southern Region of Chersonesos

D. Yu. Ponomaryov and V. P. Smirnov

This study was intended not to only determine the biological characteristics of each of the buried persons (sex, age, race, and diseases during lifetime), but also to analyze socially predetermined characteristics, such as aggregate period of life or mortality within different age groups. It is known that an individual's physical characteristics are determined genetically, but are also influenced by the conditions of his or her life, climate, nutrition, labor performed, and disease. We have succeeded in discovering some of the above-mentioned factors in the study of this sample. The analyses will continue and the sample size will increase significantly in the coming years.

Burials in Tomb I, USCHPA 164 (under the floor of a church) had been disturbed more than once, and the disarticulated bones of earlier burials were mixed with articulated skeletons both across and down (Fig. 1). As became obvious after cleaning of a set of skulls at the western wall of the grave (Fig. 2), disarticulated bones had been placed without orientation, probably to save space.

The physical anthropological materials are divided into two large series: 1) disarticulated (mixed) bones; and 2) fragments of articulated skeletons (skeletons 1–11). Besides that, the second series includes bones of fragments of two skeletons (no. 12 new and 13 new), which, as has been proved in the course of our investigation, belonged to two disarticulated skeletons. The state of preservation of the materials is satisfactory in general, though many of the bones are fragmented. Initially, bones of the first series were sorted out into groups of bones of the same anatomical description, then divided into left and right sides. One more group consisted of fragments of bones that cannot be classified as left or right. All the above-mentioned groups were divided into an adult and a juvenile series. Next we compared morphometric and morphoscopic characteristics of the bones for determination of the number of buried individuals. This study indicates that the mortal remains of over sixty individuals were deposited in Tomb I, which was in effect a charnel pit. It should be pointed out that remains of male individuals, mature or senile, predominate among the articulated skeletons.

The sex and age group of the individuals with articulated skeletons are:

No. 1: adult male
No. 2: senile female
No. 3: adult or senile male
No. 4: senile male
No. 5: adult male
No. 6: adult male;
No. 7: juvenile I (12–13 years)
No. 8: adult male
No. 9: adult male

Figure 1. Burials in Tomb 1, under the floor of a chapel.

Figure 2. Skulls in Tomb I, western wall.

No. 10: adult male;
No. 11: juvenile I (6–7 years)
No. 12 new: juvenile I (4–5 years)
No. 13 new: juvenile II (8–10 years)

Determination of a Number of Persons Buried in Tomb I
Ulna, radius, and sacrum bones are diagnostically important for determination of the number of adult individuals. The number of individuals of less than 14 years has been determined by the measurement of long tubular bones. There are also four bones of skeletons of at least three fetuses. The total number makes up 62 individuals: fetuses, 3; infants & juveniles I and II (0–14 years), 27; adult individuals (adult and senile), 32. The gender structure that emerges from this detailed study of anatomical elements reveals that the sample consisted of 56% females, 44% males. Of particular interest is the high rate of infant mortality, distributed unevenly, as might be expected. Having survived the first year, a child's chances of survival increased. This high rate of infant mortality is paralleled in other parts of the ancient (*e.g.* Metaponto), medieval, and underdeveloped and poor areas of the modern world (*e.g.* early 20th century Appalachia.)

Structure of Juvenile Mortality
(including antenatal mortality)

Age (years)	Number	Percentage
Foetus	3	9.7
0–1	10	32.3
1–7	13	41.9
7–14	5	16.1
Total:	**31**	**100**

Structure of Mortality of the Adult Population
for both male and female populations

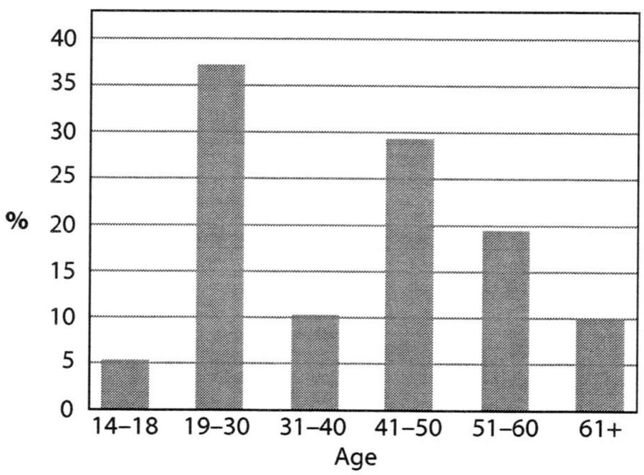

The age of adults and those between 14 and 18 years has been determined according to the state of obliteration of sutures of skull vaults and obliteration of *synchondrosis spheno-occipitalis.* We present here the preliminary data regarding the structure of mortality of the adult population based only on skull data.

Brief preliminary conclusions
•Palaeoanthropological analysis was possible for 4,821 skeleton bones from the Tomb I.

•Part of the skeleton bones were articulated (skeletons 1–11, 12 new, and 13 new). These bones were from a senile woman, 4 juveniles (4–5, 6–7, 8–10, and 12–13 years old), and 8 adult men (*adultus matures - senilis*).

•The analysis of both series of articulated and disarticulated bones together proves these were the bones of skeletons of at least 62 individuals, namely 3 fetuses, 27 individuals 0–14 years old, and 32 adults.

•The analysis of the bones from juvenile skeletons demonstrates that the aggregate age of death of individuals 0–14 years (not including antenatal mortality) is 3.21–4.06 years.

•The gender structure of the burial population is typical of ancient and medieval periods where male individuals predominate: males, 56%, females, 44%, for individuals older than 14 years.

•The age of death for the individuals older than 14 years is determined according to the state of obliteration of sutures of skull vaults (taking into account both fragmented and complete skulls).

•The diagram of the age of death of the part of the population under analysis has two peaks: 18–30 and 40–50 years. Again, this is typical for classical and medieval stable populations (with little or no immigration or emigration).

•The average age of death of male and female individuals older than 14 years old is 32–43 years. The given index of mortality of the adult and juvenile segment of the population is typical for European populations throughout antiquity and Middle Ages.

•The analysis of the skeleton bones from Tomb I have shown that the state of preservation of small bones of skeleton (bones of foot, hand, etc.) is less than 50%, which fact allows us to hypothesize that Tomb I was a secondary burial of those individuals whose bones were disarticulated.

•The finds of few articulated bones when the skeleton in general is disarticulated supply evidence that the later burials were made when the soft tissues of the previous burials partly remained.

•The evidence indicates a high degree of genetic homogeneity in this population. The study has shown the high frequency of occurrence of incomplete obliteration of the front suture (metopism) and supplementary sacrum-illiacus joint. Odontoglyphics and odontometry provide further evidence.

•The physical characteristics of the population, as a whole, are much closer to the Mediterranean-Balkan than Asian.

•The preliminary palaeoanthropological analysis of skeletal material indicates the presence of the following diseases in this sample of the late medieval population of Chersonesos: degenerative-destructive diseases (deformative arthrosis, spondyloarthrosis, deformative spondylosis, intervertebral osteochondrosis, ineterspinal arthrosis, and Forestier's disease); infectious disease (tuberculosis, syphilis, and unspecific infectious diseases of lungs); Marie-Bumberger's syndrome, diseases of blood system (*cribra orbitalia*, hemoglobinopathia, and anemia); tumors; osteoma of the left parietal bone; malignant osteosarcoma of thoracic vertebra, and osteoma of the breastbone; compensatory reactions (periostitis of tarsal bones); traumas and its consequences (breaks of fibula and sacrum, compressive breaks of thoracic vertebrae, broken ribs and nose bones, periodontal trauma, traumas of dental hard tissue, a cut on the diaphysis of femur, post-traumatic fibrositis of the humerus and an acromegalic outgrowth of shoulder bone); carious and not-carious defects of hard tissues of the teeth, periodontal diseases, and various congenital anomalies of development. Despite these many and varied ailments and traumas, the population of the Southern side of Chersonesos lived relatively long lives. This is a subject to which we shall return as our study and the excavation proceed.

Conservation of the Cistern Complex of Chersonesos

Larissa Sedikova

Archaeological investigations in the southern region of Chersonesos, begun at the end of the 20th century by K. Kosciuszko-Valuzhinich, were continued by the State History Museum under the direction of N. Piatysheva in 1950–70s. After a long interval, excavations of the cistern complex were resumed by the author in 1987. The resulting architectural complex, with the ancient city's main cistern as its central focus, is now uncovered in an area adjacent to the southern defensive wall. This complex contained public buildings, one of which, situated to the west of the cistern, was defined by the first excavators as *termae*, or baths. The interpretation of this building is arguable; however, it was doubtlessly a monumental structure that was built in the best tradition of Roman construction. It consisted of three large halls and two small rooms that were probably water storage facilities. Archaeological material allows a date of not earlier than the third century AD.

The monumental walls were located on solid foundations, which in some places were four meters deep. The walls were made of hewn blocks alternated with plinths in rows of four and fixed with lime-cement mortar. Such a proportion of mortar, which consists of lime, sand, and ground ceramic, is also called "Roman concrete" due to its solidity. In Chersonesos, as in other towns in the Roman sphere, this mortar was used to coat fish-salting cisterns. This kind of brickwork was later used in Chersonesos during the construction of monumental buildings, and specifically early Christian basilicas. In addition, vertical ceramic pipes were inserted into some walls, which probably channelled rainwater from the roof into the cistern.

The building's floors were covered with a monolithic priming of pinkish concrete mortar mixed with gravel. The Chersonesos craftsmen probably followed the recommendations of the Roman architect, Vitruvius, when laying the floor. Thus, to smooth the surface before the floor priming, a layer of quarried stone was initially laid, followed by a gravel layer, and only after this the entire surface was covered with the lime-cement mortar. The construction method of the floor is evident in areas of damage.

Figure 1. Wall of Room 3 as it appeared at the start of the 2002 season.

Although this complex is of great scientific value and one of the most enchanting sights in Chersonesos, not much attention has been paid to its conservation. Work done in 2002 aimed to preserve surviving ancient architectural traces from further destruction and to develop the most acceptable methods of conservation for this object.

Room 3, the most damaged area of this complex, was chosen for initial conservation measures. Its total area measures about 100 square meters (Fig. 1).

As the excavated monument had been in continuous exposure to the elements for a prolonged period of time without proper conservation maintenance, the effects of extensive weathering are evident. The walls and floors of the structure were gradually overgrown by grass and shrubs with their associated root systems. All attempts to eliminate the vegetation mechanically led to the inevitable destruction of the ancient structures where plants took root.

An herbicide solution was applied to the entire space of one room, including the walls, thus preventing vegetative growth for several years.

Precipitation and seasonal changes in temperature have led to freeze-thaw processes, which have damaged the structures. For example, before conservation the floor surface in Room 3 was a porous pink layer with loose pebbles, having lost its adhesive quality. We followed the recommendations of visiting specialists from Cleere Conservation Limited, UK, and covered the floor with a thin polymer netting until such time as an appropriate hydrophobic mortar is devised. The netting was then covered by a layer of fine gravel similar to the original floors, but different in composition. Such a protective covering can easily be removed and replaced. (Fig. 2)

In addition to nature, another factor contributing to the destruction of ancient walls are the visitors who freely climb them. The most vulnerable walls, those

Figure 2. Room 3 wall after conservation.

most accessible to visitors, have been temporarily covered with sandbags. This measure, while not aesthetic, is the only way to preserve the ancient masonry until a proper method of conservation is decided upon. (Fig. 3)

Lime-sand mortar mixed with ground ceramics, a composition closest to the ancient mortar, was used to conserve a small experimental area of the western wall of Room 3.

Samples were taken from the walls in the complex in order to study the composition of the ancient mortar. Laboratory analysis of samples was conducted at the Construction Administration of the Black Sea Fleet. The mass of the mortar samples was 12.1 kg/sq cm. The proportions of the mortar were 70% lime, 30% fine sand and granulated ceramic material fill.

Further observation of the experimental wall will be conducted in order to ascertain the adequacy of the mortar for the conservation of other walls. The initial results of the experiment will be analysed in spring 2003, after the seasonal change in weather. If the results are positive, we will be able to use this method for further conservation of the entire architectural ensemble in the cistern area, after which the complex can be included in the excursion route of the ancient city.

Figure 3. Visitors are discouraged from climbing on fragile walls by a mass of sandbags—a temporary solution until the area can be properly conserved.

DOCUMENTING THE MOSAIC FLOOR OF A BYZANTINE CHURCH IN THE ITALIAN GARDEN

Adela Sobotkova
Masaryk University of Brno, Czech Republic

In the summer of 2002, during the excavations in the city and chora of Chersonesos, we started a project to photographically document a mosaic floor in the Museum Courtyard.

Objectives of the project
The goal of this photographic project was to record the present state of the mosaic floor which comes from a Byzantine cross-shaped basilica in Quarantine Bay. The basilica, which was erected over the shrine of the martyr St. Martin, dating between the last quarter of the sixth to the eighth century AD, was found well-preserved during the excavations in 1902. Its unearthed walls run 3 meters high and some of the window rims are still visible. During the Second World War the church was damaged by a bomb; the mosaic floor suffered from fire and had to be cleaned and reconstructed during excavations in 1951–53. This floor is now situated outdoors in the Museum courtyard, also called the Italian garden (the official name being Greek Garden), on the grounds of the Preserve. Since its relocation it has been weathered by climatic changes. The damage is most obvious along the edges, where individual *tesserae* gradually loosened and broke off. Parts of the surface are covered by moss and the colors have become shrouded. The easy accessibility made it also possible for negligent visitors to dismantle and carry loose pieces away to satisfy their desire for souvenirs. Because of the deteriorating condition of this remarkable monument, we decided to create a digital record for study and future preservation.

Methods of Photography
When deciding on the methods of photographint the mosiac, we had to deal with several difficulties, such as its large size and narrow space to manuever. Another challenge was assuring even lighting of the surface and choosing the right procedure in order to get well colored, sharp and rectified pictures, which could be easily processed and stitched on computer. We wanted to make use of the advantages of digital photography, such as fast advance and immediate feedback in the field, with the possibility of correction and flexibility during computer processing.

The mosaic spreads over the T-shaped floor retained from the original cross-shaped church floor. The dimensions of the main nave are 4.94 by 12 meters and of the transept 15.13 by 5.08 meters. Having such a large subject and being limited in time, we decided to photograph only the central part of the mosaic that spreads over the crossing. Its area was still rather large, 6 by 5 meters; therefore to get well rectified pictures we needed to take them from an appropriate height, possibly using a high ladder, so that each image covered a larger surface with less distortion. However this was out of question because surrounding vegetation reached far and hindered the view from higher posts. Our solution was to use a 1.6 m tall and 3 m long construction made from metal tubes. We fastened the camera on the horizontal bar supported by two poles at the edges and slid it along to take pictures at equal distances from each other, as well as from the surface of the mosaic. To assure that the pictures were taken from the same angle, we leveled the camera before each shot. For easier orientation on the large surface we drew up a grid and marked it on the mosaic. We moved the construction along it fully covered the mosaic with a sequence of successive images.

To achieve even lighting of the mosaic we decided to take pictures at night and use artificial light. At night we could avoid uneven daylight, as well as the permanent shadows created by the branches of the trees in the immediate surrounding. Originally we intended to light the surface with studio lights, but an insufficient electrical supply restricted us to using flash. This seemed to be a good choice, but problems showed up

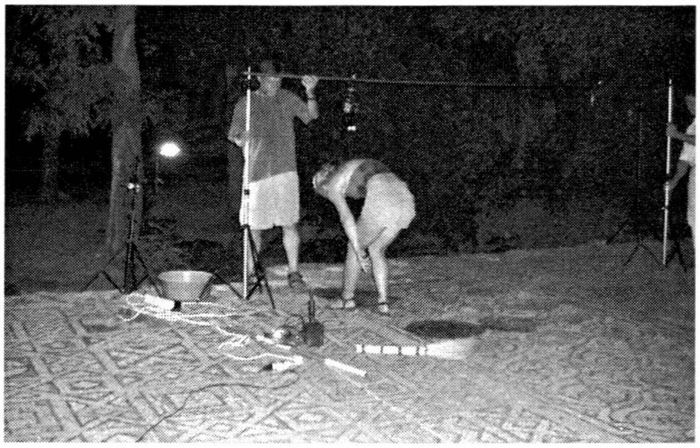

Figure 1. Photographing the mosaic in the Italian Garden.

when the images were processed at the computer.

In order to enhance the color, the mosaic was swept clean and lightly moistened with water before photographing. The short distance between the camera and the surface demanded the exposures to be taken every 1 m in length and 0.5m in width. Actual area covered by the lens during the exposure was larger, but we had to include a reserve for satisfactory overlap of individual images. During our session we documented the area of the crossing which covers 6 x 5 meters. The composite at the end constituted of total number of 58 images.

We used a Canon Power Shot G1 digital camera for taking the pictures. The biggest advantage of digital camera is the amount of data that can be recorded, transported, and quickly processed on the computer. In the field its exposures can be checked and corrected on the LED display and, in case of need, reshot on the spot. The resolution for the images can be set according to the conditions and needs. We started from the upper left corner, moving our camera support in a linear pattern along the chosen part of mosaic, taking around 100 shots at a resolution of 2048 x 1536 pixels. The photography was done in two sessions (with a break for battery exchange) under the direction of photographer Chris Williams, assisted by Richard Posamentir and the author.

Computer Processing

The images were processed in Photoshop and Canon Photostitch program. Their file size was decreased and unified. During the stitching some of the weaknesses of our methods emerged.

To our surprise the zoom changed after a certain number of shots, which was not noticeable in the view provided by display. The cause might have been the characteristics of the G1 to reset to the wide angle when sleeping or turned off. The greatest variation in zoom showed in the middle of the mosaic, which corresponded to the time of battery exchange. Nevertheless, variations of the zoom were corrected during computer processing.

The flash sometimes created reflections on the wet surface and often affected the brightness of the images. Different color intensity became obvious in the burnt-in edges that formed a grid of darker lines on the composite after the stitching. This unpleasant noise had to be corrected by the tools in Photoshop. Next time it could be prevented by providing more even tungsten lighting of the surface. Another unpleasant surprise was the fact that the automatic focus of the G1 Power Shot camera was not always consistent. This was caused by low light levels that misled the auto-focus indicator. This deficiency could be prevented by using an upgraded digital camera. However, for the purposes of this initial publication the sharpness is satisfactory as the composite has been shrunk to a smaller format.

Some images were slightly tilted from the designed grid scheme, due to the lack of precision of our support apparatus. Because the images lacked enough overlap area, they were rejected by the Canon Photostitch program and had to be attached manually in Photoshop. The difference between the automatic and manual attachment can be seen below. The Canon Photostitch program automatically produces a well-rectified picture after it has been fed good quality images, proper setting, and directions. The limitation of this program is the fact that it is rather sensitive to the properties of the images and requires that they are not tilted, are unified in size, and have a large overlap area. It produces good composites, but it is not flexible enough and does not have enough tools for manipulating the images. The rejected exposures must be processed beforehand in different program. This disables immediate comparison and mutual alignment of all the pictures in a row, making the task rather chaotic and laborious. Due to the number and uneven quality of the images, as well as the final size of the composite picture, Adobe Photoshop turned out to be the more effective program. Although it took more time and energy to put all the pictures together manually, its numerous tools and flexibility assured a good final outcome; therefore, in the end all the work was done in Photoshop.

Conclusion

Our project of documenting the mosaic with the use of digital photography turned out to be an interesting challenge. We had to deal with difficult photographic circumstances. We took and processed a large amount of pictures and became aware of the limitations of the digital camera as well as the available software. However, in the end we acquired a satisfactory result that can serve as an incentive for next the photographic projects as well as for study and preservation purposes.

Mosaic description

The mosaic adorns the T-shaped floor of what originally was the interior of a cross- shaped Byzantine basilica that served as a memorial church of St. Martin. It consisted of a main nave with a transept. The eastern extension of the cross, the choir, has not endured till present; its floor had not been covered with the mosaic, but with marble slabs. The mosaic therefore had the shape of T originally. The central part of the mosaic floor is well preserved, except for a few holes in the lower part of the square *emblema*. In peripheral areas, however, many *tesserae* are missing or loose and crumbling away. (See Plate 1, p. v.)

The mosaic is made up of black and white, as well as brightly colored, evenly cut square *tesserae*. The prevailing colors are black, white, yellow and red. The white tiles are made of marble, black of a special sort of sandstone from Balaklava, red of limestone or pieces of brick ware, and yellow of locally exploited limestone. There are also some blues, made of glass cubes.

The central part of the mosaic shows a colorful motif of a vase with two peacocks at the sides, surrounded by vine tendrils with grapes and little birds. The vase represents a wine drinking vessel (*kantharos),* a frequent Christian symbol of eternal life and eucharist. Its shape is a creation of an artist, highly decorative but impractical. It has a white wavy line on black rim, wide opening, funnel-shaped neck, oval shoulders, embossed body and a long slim leg. The volume of the embossed body surface is nicely suggested by color gradation. In addition to the optical illusion of volume created by shading, the same effect is achieved by the *alinamenti* of *tesserae* that follow the shapes of the depicted objects. Extremely thin handles with volutes at the ends touch the *kantharos* at the rim and at shoulders. The neck is decorated by three ribs. In the lower part of the mosaic the leg is damaged and missing. Otherwise the whole vessel is gracefully shaped and well proportioned.

The wine branches emerging from the *kantharos* turn symmetrically in opposite directions and surround the vessel in rhythmical waves, ending above the backs of the peacocks. Their undulations are filled with single leaves, grapes and two birds. The branches are brown, as are the leaves and the birds.

The peacocks are situated in a semi heraldic position on the sides of the vessel. They are slightly smaller in scale than the vessel and their heads reach only to the widest part of its belly. They are rather schematic, strongly outlined, and can be easily recognized by the shape of body, crowns and richness of color. They are not very detailed. They have long yellow green necks, red breasts, blue green backs, and white bellies. Their wide tails are simplified as the feathers are only schematically indicated by lines in black, red, and yellow. The peacock occurred as a frequent motif in antiquity where he symbolized immortality, because its meat was considered imperishable. Christian iconography adopted the bird as a symbol of immortality and resurrection. As such it appears very frequently on Christian mosaics throughout the Byzantine Empire.

The central motif, enclosed in a white square frame, is surrounded by a network of colorful medals that show representations of animals, tools, and symbols of Christianity. There are two white stripes separating the motif of the crossing from those in the transept. The medals are aligned in a set pattern of rows on the sides of the central *emblema* and form a carpet of interlaced ornaments at regular distances from one another. The rhythmical pattern of their composition, together with disconnected depictions, lead the attention of the viewer to the central motif. A coherent scene was not the aim the artist; his main intention was decorating the floor. On the borders of the crossing the colors on some medals have been reduced to black and white. On the left we can see a transition to purely geometric black and white ornaments.

The medals contain various representations of animals and tools and symbolic objects. The majority of creatures are birds—ducks, geese and dove; others show horses, goats, ram, or fish. There are grapes and bowls of fruit, everyday tools such as drinking vessel; flowers, vine leaves, roses, representations of trees, shells, and geometric ornaments. These objects often carry symbolic meaning, e.g. the fish is a symbol of Christ and Christianity; the dove is a symbol of faithfulness and saintly spirit, the birds altogether are intended to allude to the garden of Paradise and call the thoughts of mortals to rise in spiritual spheres. The ornaments in the vicinity of the centerpiece are in full color, with strong contours, and as the *tesserae* are of the same size with those of the central large motif, the medals are less detailed and have rather flat color distribution. This resembles the traditions of Byzantine mosaic art giveing the depicted objects a simplicity and expressiveness.

The mosaic bears much resemblance in quality and decoration to its contemporary mosaic fromm the "Basilica in-Basilica" in the harbor of Chersonesos (Fig. 2) which, too, features medals with birds, geometric ornaments, and even a similar peacock motif in its aisles. They are almost identical in technique, color, and shape of animals and may have come from the same workshop. The mosaic from this Basilica has been recently removed from its place of origin to the Museum grounds for preservation.

Another analogy to the central motif can be found in the sixth century Roman Byzantine mosaic from a memorial chapel in New Jerusalem, the so-called Bird's mosaic. It shows the same motif of two peacocks on sides of a vessel with emerging vines that form aligned rows of enclosures for the birds. However, the peacocks are more detailed and proportionate, their tail feathers are depicted carefully, and their colors are more natural. The vessel is hardly distinguishable as it is overgrown with acanthus and almost transformed into a plant. Its embossed, highly decorated surface underlines its symbolic function, the source of eternal life. The composition is symmetrical, but livelier, contours are thinner, and color gradations are finer.

Although the mosaics differ in formal elaboration, the message remains the same: an allusion to the garden of Paradise

In these presented works of art, early Christian mosaicists followed the traditional techniques of Greco-Roman mosaic art. Their intention, however, was not illusory; they did not tell a coherent story but decorated the space. They used motifs that were mutually isolated but complete in their symbolic meaning. Christian iconography seems to have had a strong influence on the repertoire of Christian artists. Throughout the Byzantine Empire we can find similar motifs on Christian mosaics differing only in formal elaboration. In the mosaic from the cross-shaped church in Chersonesos, with its harmonious composition, purity of shapes and colors, we have the testimony of a highly decorative style and mosaic mastry achieved by early Christian artisans of Chersonesos.

Figure 2. The "Basilica-in-Basilica" on the shore of Chersonesos.

Ukrainian Publications and Technological Development

Glenn Mack
ICA, Austin

ICA hit full stride in 2002 with its long-term communications projects in Ukraine, which may be categorized generally into three areas: 1) publications, 2) information technology, and 3) public relations. Much of our energy and focus throughout the year was devoted to the publication of *Crimean Chersonesos*, scheduled for release in the spring of 2003. The richly illustrated volume is an extensive introduction to the history, archaeology, and monuments of the National Preserve of Tauric Chersonesos. We also implemented the colossal technology project, dubbed *Megarika*, to digitally preserve materials from the Preserve archives, library, and fondi and place the information in a unified database. Finally, ICA and the Preserve received considerable attention in the media this year, including a cover article in *Archaeology Magazine* and a feature television program entitled *Charge of the Light Brigade*. As part of the British series *Battlefield Detectives*, Grenada TV documented ICA's archaeological and technological work at Bezymyannaya to cull clues from the time of the Crimean War (1853-56), as the site was used as a defensive position by Turkish troops.

Figure 1. The November–December issue of AIA's *Archaeology* magazine featured Chersonesos in its cover story.

Publications
ICA is honored to be producing the first English-language description of the National Preserve—ancient city, chora, museum, and surrounding historic sites of southwest Crimea. *Crimean Chersonesos* was produced with over four-dozen collaborators from around the world, bringing the richness of the site to a Western audience. Documentation of the site is particularly crucial as the Preserve applies for UNESCO's World Heritage List. We also laid the groundwork to produce and publish three English-language archaeological journals with Ukrainian collaborators. The first publication, *Chersonesos Studies*, is a joint project with the National Preserve and ICA, bringing to light the most significant work on the ancient city and its chora. The second journal reviews the latest developments in Ukrainian archaeology and will be co-produced with the University of Kyiv Mohila Academy. ICA is also publishing a joint year-end volume of selected articles from *Arkheologiya*, the National Academy of Science's Institute of Archaeology quarterly publication devoted to ancient history, archaeology, and excavation results in Ukraine.

Information Technology
In 2002, ICA directed the development of *Megarika* and the museum website. We fully implemented the *Megarika* database project with the hiring of key computer programmers, the creation of a unified database, and the installation of essential equipment. National Preserve Head Librarian Lucy Grinenko is the on-site coordinator of *Megarika*. We were greatly saddened in the summer of 2002 when Oleg Vertolitsky, designer of the database and website, tragically died. His enthusiasm and brilliance will be sorely missed. By September, we hired three of his former classmates as the core *Megarika* team. Aleksander Semenov serves as the technical leader, Andrei Kondyuk digitizes the material, and Aleksei Romanov is the systems administrator. After several weeks of testing the equipment and integrating the software, the *Megarika* team spent the remainder of the year refining the database and the parameters for data storage. They consulted with the department heads and provided training to museum staff so they may assist in data acquisition and entry. The team tested the database with a variety of formats: illustrations, photos, field notes, and inventory cards from the archives, fondi, and library. They began to scan and digitally preserve the most delicate materials for storage and retrieval by scholars, specifically the complete field notes of the first museum director Karl Kostsyushko-Valyuzhinich, among the most valuable and accessed documents of the Archive. A prototype database version was also tested for restricted access via the Internet. In addition, Lucy Grinenko, Taissa Bushnell, Svetlana Telenkova, and Nikita Khrapunov continued their development of the *chersonesos.org* website with information about the museum and its archaeology in Russian, English, and Ukrainian. The site received high praise from the Ukrainian Ministry of Culture and Arts when it was formally presented in June 2002 in Kyiv.

Figure 2. Glenn Mack at the Kyiv airport waits to load the document scanner into Vasili Kirichenko's truck for the trip to Sevastopol.

Public Relations

In 2002, ICA assisted in the production of two major media works and hosted Lucy Grinenko in a scholarly exchange. During the summer season in Chersonesos, a UK production company Granada TV, filmed a documentary in Sevastopol as part of their *Battlefield Detective* series. The *Charge of the Light Brigade* relied heavily on ICA expertise, including the use of technology in archaeological research and the results of our excavations at Bezymyannaya, a site which saw action during the Crimean War. ICA's Jessica Trelogan spoke at length on camera about her data interpretation of satellite imagery, and former Bezymyannaya field director Phil Freeman shared his knowledge of military history of the site and the surrounding area. The series will be distributed worldwide in 2003. *Archaeology Magazine*, a publication of the American Institute of Archaeology, carried a cover story on Chersonesos in their November 2002 issue. ICA assisted the publication with photographic materials and fact checking in an article that focused on the successful Church-State relations on Preserve territory. Ms. Grinenko also wrote a separate article on the history of the Chersonesos Museum. Fortuitously, Lucy was in the US for the entire fall semester as part of the "Contemporary Issues" IREX program. She made several public presentations about Chersonesos and used her time at the university to research methods for building and sustaining digital collections, specifically investigating models for libraries, archives, and museums.

Figure 3. Staff of the Archives and members of the Megarika project gather as the document scanner is put into service.

PIDTRYMKA CHERSONESU (SUPPORT FOR CHERSONESOS)

Taissa Bushnell
Director, Pidtrymka Chersonesu

The goal of the non-profit organization Pidtrymka Chersonesu (PC) is the preservation of cultural heritage located within the National Preserve of Tauric Chersonesos (NPTC). PC was created by the Institute of Classical Archaeology and registered in Sevastopol in the summer of 2001 to address the need for fiscal regulation, a need created by the expansion of the activities of ICA at the Preserve. The National Preserve, under the jurisdiction of the Ministry of Culture and Arts of Ukraine, received almost no financing from Kyiv, except for staff salaries, for all of 2002. Pidtrymka Chersonesu, funded through the efforts of ICA, supplements the Preserve's budget for projects deemed necessary by joint decisions of the General Director of the Preserve, Leonid Marchenko, and Professor Carter, and it works closely with all levels of the staff during the course of the year to assess its needs and review the effectiveness of its financial support.

Tools for Lifting Barriers

In 2002 PC continued financing some of those activities and services started in 2001, such as computer training for staff, Internet access, and the conservation of Site 132 in the chora of Chersonesos as part of the development of the planned archaeological park. The spirit behind PC's program, to a large degree, is to ensure that the Preserve staff feels the direct benefits of its various activities. The Internet, for example, has not only enabled the easy and unbroken connection between the Institute of Classical Archaeology and the Preserve during the entire year, but it has also provided a link with other museums and universities in and outside of Ukraine, and has facilitated exchange between scholars at Chersonesos and contacts formed at international conferences.

With PC's assistance, Lucy Grinenko, head librarian at the Preserve, attended a conference on library digitization in St. Petersburg in May 2002. Oleg Vertolitsky, employed by PC as systems administrator for the Preserve museum, returned especially invigorated from a conference in Nizhny Novgorod in June with the affirmation that his work on Megarika, the data base project involving the digitization of the Preserve's archive, stores, and library, was on par with the most advanced projects discussed. Oleg tragically died in July. He will be missed by all of us and his work on Megarika will be a lasting reminder of his vision for the project and its possibilities.

PC sent two particularly dedicated students from this summer's ICA-NPTC-Lecce University joint excavation, Nikita Khrapunov and Danila Kostromichov, from the Institute of Oriental Studies at Tauric University in Simferopol, Crimea, to two conferences this year: "Archaeology and Ethnology of Eastern Europe: Youth Steps Forward into the 21st Century" in Odessa, and an archaeological congress in Kharkiv.

The availability of the Internet and the participation at conferences of Preserve staff and students whose interest and work in Chersonesos promise to continue well into the future are ways in which the Preserve has enhanced its contact with the international community, contact which until recent years has been hampered, indeed rendered almost impossible, by the "closed" status of the city of Sevastopol.

Figure 1. Semih Ercan and Chris Cleere examining conservation work at Site 132 in the chora.

Likewise, the Institute of Classical Archaeology is helping to bring international attention to Chersonesos through the publication of the upcoming *Crimean Chersonesos*, and also by supporting the publication of part one of a series of informational pamphlets about the exhibits of the Preserve's museum. This first installment, "Byzantine Steatite Icons," has been financed by PC, and is available in English, German, Russian, and Ukrainian.

Nomination to the World Heritage List

During the summer, Professor Carter invited Dr. Henry Cleere, World Heritage Coordinator of ICOMOS (The International Council on Monuments and Sites), to assess the state of conservation and management of Chersonesos and its chora and to provide suggestions for improvement of the site in view of its imminent nomination to the UNESCO World Heritage List. At a meeting with Leonid Zhunko, mayor of Sevastopol, and members of the Sevastopol City Council, Dr. Cleere spoke of the importance of Chersonesos, its potential inscription on the World Heritage List, the benefits that inclusion could bring to the city of Sevastopol, and the responsibility of the city to maintain its unique cultural heritage. Professor Carter showed Mayor Zhunko a plan for the continued development of the chora of Chersonesos and stressed the danger of urban encroachment on ancient sites. Jessica Trelogan's time series of satellite images tracing the growth of urban development over the past two decades made a particularly profound impression.

The Preserve is currently implementing Dr. Cleere's recommendations and has begun to take steps to rectify obstacles standing in the way of its inscription on the World Heritage List. In May 2003, a UNESCO delegation will be visiting several Ukrainian sites that are being considered for nomination. The nomination process relies heavily on the participation of the Preserve staff because of their substantial knowledge of the site and the fact that they are one of the principle stakeholders of the site. In consideration of this, the author will be making a presentation to the Preserve in January 2003 about UNESCO, the World Heritage List, the reasons for nominating Chersonesos, and what the nomination process entails. The author's attendance at the "Space Applications for Heritage Conservation" conference in Strasbourg in October and at the ICOMOS General Assembly in Madrid in December has also provided insight into heritage conservation issues facing other sites around the world and into the steps for inscription to the World Heritage List, knowledge which will be shared with the Preserve. Assistance in nominating Chersonesos to the World Heritage List will be one of the main aims of Pidtrymka Chersonesu in 2003.

Conservation

One of the programs of Pidtrymka Chersonesu is to act as an advisory body to the Preserve in heritage conservation methods that adhere to international standards. Before nomination to the World Heritage List, a conservation plan for Chersonesos must be in place. Under a grant provided by the Global Heritage Fund, two specialists in the field of stone conservation, Chris Cleere of Cleere Conservation Limited, and Semih Ercan, were invited in August to review the state of conservation of Chersonesos and its chora.

Part of the grant, administered by Pidtrymka Chersonesu, was used to begin conservation of a section of the ancient city, the western structure of the cistern complex, which was in particularly bad repair. Work by a local firm, Krymrestauratsiya, was directed by Larissa Sedikova, deputy director in charge of conservation at NPTC, and Stanislav Ryzhov, head of the department of architecture and archaeology, and was based on the recommendations of Mr. Cleere and Mr. Ercan.

Plinths used in the Roman *opus mixtum* method of construction (levels of brick plinths interspersed within a stone wall) were crumbling away from some walls within this structure. These walls were covered with netting and sand bags in the interim while a proper mortar could be devised. Another wall was chosen for the application of an experimental mortar, the content of which was based on laboratory analysis of the original mortar. Vigilant observation of the new mortar (sand, lime, crushed ceramic) will detect how well it withstands the seasonal changes in temperature, after which a decision will be made whether it should be applied to other walls.

Rehabilitation of Storage and Laboratory Building

With the generous support of the Packard Humanities Institute, an ambitious construction project has been launched, which will rehabilitate a 19th century building, originally part of the monastery complex that once existed on the territory of Chersonesos. The new structure, while retaining its historical appearance, will house modern storage facilities for archaeological collections, a finds laboratory, and work spaces.

Initial planning meetings in May 2002 involved members of the Preserve, PC, Lex Service (Sevastopol law and auditing firm acting on behalf of PC), and Ludmila Dukareva, architect and engineer licensed in Sevastopol. Together, all parties reviewed architectural and engineering specifications, which were created in 2001 by Llano Design (Austin, Texas) in collaboration with Nikolai Andrushchenko from the State Institute of Urban Architecture (Kyiv), and discussed in detail how the plans translated into reality onsite.

Thorough review of the building by local city building inspectors followed. They required a geologic study to ascertain the viability of the material under the building's foundation. GeoKom-Intiz (A. Malovany, director) undertook the analysis of cores. Upon receipt of the geologic study, the inspectors gave their approval for construction and their recommendations.

The arduous task of finding a temporary home for the artifacts located in the building, including large ancient Greek grave *stelai*, then followed. In July ICA sponsored and supervised the move of the *stelai* into the Ancient Hall of the museum. An official agreement was signed in August for the drawing up of architectural, engineering, and technical plans based on those of Llano Design, and which would adhere to local Ukrainian law and regulations. Construction is planned for 2003.

Outreach

Pidtrymka Chersonesu has maintained regular contact with the U.S. embassy in Kyiv through meetings with Ambassador Carlos Pascual, Public Affairs Officer Janet Demiray, Cultural Affairs Officer Paul Hacker, and Assistant Cultural Affairs Officer Marta Pereyma in order to update them on the progress of our many projects at the Preserve and of ICA's support of Ukrainian student participation in summer excavations. The embassy has been very helpful as a source of advice on student exchanges and grants available through the U.S. Department of State. We are indebted to the ambassador for his continued support of ICA's projects at Chersonesos and for his assistance with the Ukrainian government.

The support of the Ministry of Culture and Arts of Ukraine is integral to the success of ICA and PC's activities at Chersonesos, especially of those projects requiring governmental approval, like the reconstruction of the storage and laboratory building. Professor Carter's meetings with Minister Yuri Bohutsky, and the author's meetings with Evnika Linyova, head of the Department of Cultural Heritage at the Ministry, have been instrumental in securing the Ministry's aid on a number of occasions.

In February 2002, discussions with Dr. Leonid Zaliznyak, director of the Master's program in archaeology at Kyiv Mohyla Academy, were fruitful in laying the groundwork for long-term collaboration with ICA. The author's subsequent lecture to undergraduate and graduate archaeology students detailed ICA's history at Chersonesos. Twelve students from the program joined the summer 2002 excavation teams, and an agreement was signed in June by Professor Carter and the Academy for further scientific collaboration between the two institutions, including annual student participation at ICA's excavations in Chersonesos and a jointly published archaeological journal.

Figure 2. Taissa Bushnell and Ambassador Carlos Pascual at the U.S. embassy's Fourth of July festivities.

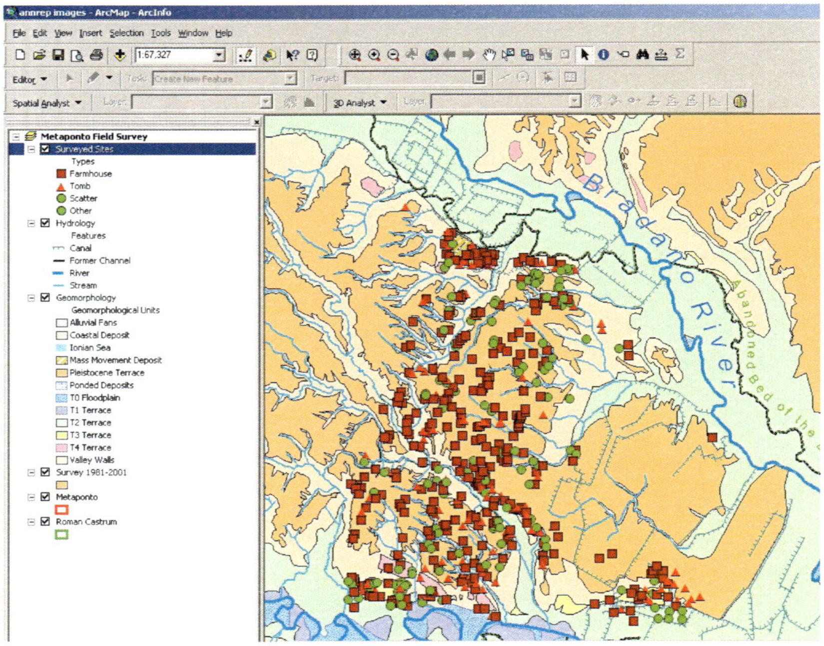

Plate 10a. Screenshot of Metaponto GIS data in ArcGIS 8.2. (See p. 47.)

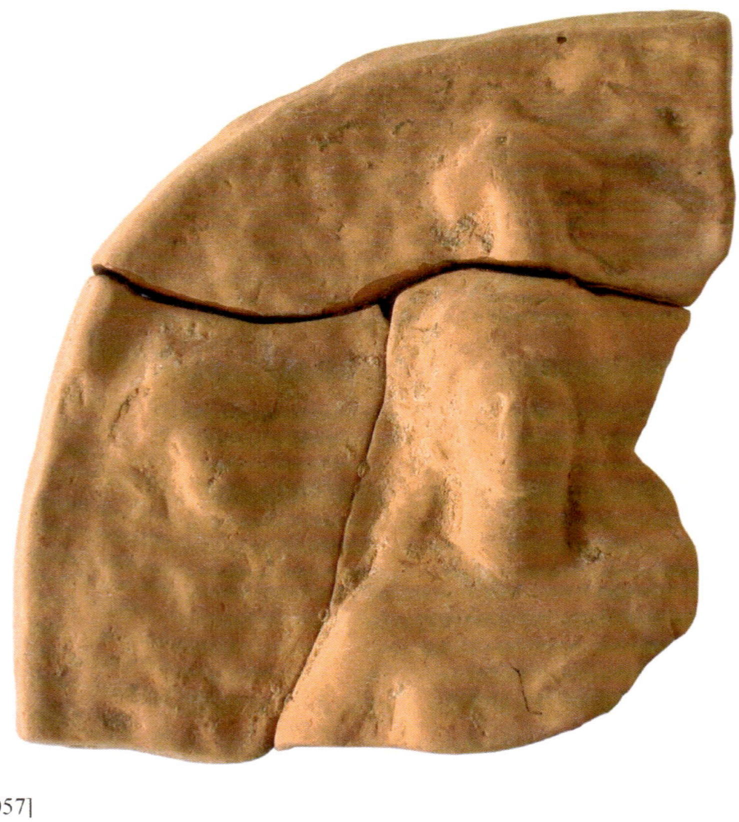

PZ 77-933 [321057]

PHOTO: GIAMBATTISTA SASSI

Plate 10b. Terracotta plaque from the Pantanello Sanctuary representing a maenad. (Photo: G. Sassi) (See p. 63.)

Intensive Field Survey Between Bradano and Basento: Publication, Progress, and Perspective

Alberto Prieto
University of Texas at Austin/ICA

GIS of the Metapontine Chora

ICA's Geographic Information System (GIS) covering the areas of the Metapontine chora (or Metapontino) studied through intensive field survey (1981–2001) is the constantly-evolving digital platform upon which all research and publication efforts have been built since the resumption of survey activity in 1999, after a five-year hiatus. GIS has become ICA's primary tool in the study of the human and natural landscape at both Metaponto and Chersonesos, and it will doubtless retain this position for years to come. GIS is the subject of increasingly lengthy sections in recent *Annual Reports* (1999–2001), with perhaps less description of its functions and organization than many readers would like. With major re-working of the Metaponto GIS in 2002 (see below), this is a fitting occasion to introduce the topic in greater detail so that our audience will begin to appreciate the central importance of GIS in ICA's mission and results.

In the most essential terms, a GIS is a software package, running on a personal computer, that handles any type of geographically-referenced information (i.e., information about, or associated with, a geographic location). A GIS is a digital recreation of the real physical world—a necessarily simplified geometric model of highly complex geography—in which any given object (feature) can be described in terms of its absolute position (location), its physical relationship with other features (distance, direction, superposition, etc.), and its attributes (size, color, kind, etc.). Humans perceive and organize the physical world in two components: visual (appearance) and cognitive (description). To mimic human experience, a GIS program combines high-resolution graphical rendering (which displays different classes of geographic data using different colors, sizes, and other visual cues) with a simple database (which organizes and stores cognitive information, such as name, kind/type, use, and so on). Data in a GIS can be manipulated—sorted, selected, analyzed, and so on—either visually, via the graphical interface (for example, by selecting all sites within a 1km radius of a given point), or descriptively, via the row-and-column interface of the database (for example, by selecting all sites that are classified as "farmhouse"). Most archaeological projects, whether excavations or field surveys, have adopted GIS in the last decade, having recognized archaeological research's inherent dependence on location: in order to interpret a given site or an entire landscape, an archaeologist needs to know where archaeological features (e.g., farmsites, burials, ancient canals, or roads) lie in absolute terms, where they lie in relation to each other and to geographical features such as rivers or geomorphological formations, and how they are classified or identified.

GIS data is typically organized in datasets called "layers" or "themes," each of which describes a set of features composed of only one of the three fundamental geometric/geographic entities (point, line, and polygon); features of the same geographic entity type are distinguished by designated attributes of interest to the user. For example, rivers and streams are both represented by linear features within a GIS, but they may be stored as two separate data-layers, if the user so wishes, to make a clear distinction between them; they may also be stored in a single layer comprising all forms of waterways, perhaps along with canals and irrigation pipelines. Layers are added to, or removed from, the user interface at will, and they can be "turned on and off" and adjusted (by color, thickness, size, etc.) in the visual display. GIS is therefore an indispensable cartographic tool, capable of generating customized, descriptive maps (displaying names, boundaries, etc.) or thematic maps (displaying density, altitude/elevation, etc. using graduated color-ramps or other patterning) in minutes, saving hundreds of hours compared to traditional manual cartography. Professional GIS programs often provide a suite of statistical and analytical functions, expanding the software's capabilities to include data modelling and exploration, making it a true research tool. And no GIS is considered complete without a sophisticated export or printing interface that produces high-quality paper maps or digital images for dissemination.

The suspension of fieldwork in the chora of Metaponto in 2002 provided invaluable time for two significant upgrades to the Metaponto GIS. First and foremost, a significant change in platform was made, from the ArcView 3.x GIS software to the ArcGIS 8.x software. Though both programs are products of the same GIS research corporation (Environmental Systems Research Institute, or ESRI, of Redlands, CA) and serve the same purposes, the change is akin to replacing a map and compass with a Global Positioning System (GPS) receiver for feature-location in the field. ArcGIS represents a quantum improvement over ArcView in terms of the number and range of analytical and display functions, speed and stability of data processing and visual rendering, compatibility with other analytical software and file formats, data organization, and visualization. With ArcGIS, data is easier to store, retrieve, and manipulate, and map products, whether on paper or disk, are of much higher quality.

One of the main advantages of ArcGIS is its new "architecture," which stores multiple layers/themes of GIS data in single, industry-standard database files, rather than—as in ArcView—storing each layer in a separate file (each of which is in reality composed of from three to seven separate subfiles, increasing the risks of errors when moving data from one computer to another). A regional GIS is now much more portable, with only one file to copy or transfer. Among the most important new features in ArcGIS is its provision for "metadata," or data about data: each layer in the GIS database can be documented in terms of its date of creation, creator, proprietary entity/organization, geographic extent, source, and so on, allowing users to document and trace the histories of thematic data layers. For ICA's purposes, a wide range of fragmentary metadata left by previous researchers on paper and on disk can now be assembled in one location and digitally associated with the relevant data. Much of the summer was spent by the author in transferring the Metaponto GIS data from ArcView format to ArcGIS format and in creating the metadata for the most common and important data layers. A great deal of time and effort was also invested in learning the functions of the software in detail, in order not only to use it intelligently but also to train future users in Austin and on-site at Metaponto.

Figure 1. Eloisa Vittoria, Cesare Raho, and Vincenzo Cracolici at work in Metaponto.

The second major component of the GIS upgrade was a thorough revision of the data, searching out errors and inconsistencies introduced over the years during data-entry, as well as expanding the geographical extent of ICA's GIS data to cover more of the Metapontino. Data-layers were systematically examined and corrected, gaps contained therein were filled, and the spatial extents of many pre-existing layers were extended beyond the confines of the ICA survey transects to produce aesthetic and informative maps at smaller scales (i.e., including more of southeastern Basilicata). Additionally, 1:50,000-scale Italian national topographic maps were scanned and georeferenced as aids in locating features and relating the Metapontino to the larger regional context. The remaining folios of the Superintendency's 1:10,000-scale scale topographic maps of the Metapontino, priceless documents of the "Adamesteanu era" (1964–1977), essential to any archaeological endeavor in the chora, were also scanned and georeferenced to provide coverage of the entire archaeological zone outside the ICA survey transects. Finally, all scanned maps and data-layers were re-named using a standardized nomenclature for easy reference and retrieval within the Windows operating system.

The summer's labor has a potentially larger benefit beyond mere accuracy and completeness. ICA has proposed to the National Archaeological Museum at

Metaponto and the Superintendency of Antiquities of Basilicata an international collaborative research effort encompassing the entire Metapontino. ICA's GIS could be expanded, in both geographical and chronological terms, to document all known archaeological sites and discoveries of the entire chora, serving as a centralized research and archival tool for all parties. This would be a rare, if not unique, collaborative effort between American and Italian academic and public institutions for the preservation and study of cultural resources. The proposal is currently under review by the Italian Ministry of Culture.

Survey Publication

2002 has been a crucial year for the preparation of the second volume in the Chora of Metaponto series, Archaeological Field Survey I: Bradano to Basento. The scope of the volume has been expanded beyond its original limits (the first phase of survey between 1981 and 1990), to include all phases of ICA survey activity between the Bradano and Basento Rivers (1981–2001). This re-organization will allow ICA to present the results of the intensive field survey in terms of three distinct geomorphological and geological zones: the coastal plain/first marine terrace (1983, 1999–2000), the main body of marine terraces and the river valleys (1981–1984, 1990, 1993, 2000–2001), and the hilly zone marking the transition from terraces to mountains (1981–1982). Furthermore, the greater geographic cohesion of the volume will permit a discussion of the aims and results of the re-survey project (2000–2001), which will act as an "addendum" to the main body of survey research, documenting the changes to the landscape and to the physical condition of the sites twenty years after the beginning of ICA's intensive survey project.

The publication has also received fresh momentum from the assembly of an international team of scholars and experts on material culture, who are studying the various categories of artifacts recovered to help determine the types, functions, and dates of the sites. The most chronologically sensitive artifact type, and therefore the foundation for the chronology of the sites, is the ubiquitous black gloss pottery. Dr. Vincenzo Cracolici, an outstanding archaeologist and expert on the black gloss ceramics of Metaponto's *kerameikos* (potters' quarter), has finished a general chronological evaluation of the sites containing black gloss ceramics and is currently working on a typology of the black gloss forms, which is based on the typology developed by Maria Elliott during her study of the black gloss pottery from the Pantanello Necropolis. (See "Black Gloss Pottery from the Metaponto Field Survey," pp. 52–53.)

Dr. Cracolici has been assisted by two highly talented archaeologists and draftsmen, Cesare Raho (University of Lecce), and Eloisa Vittoria (University of Bologna), who have prepared illustrations of dozens of the most important fragments. (Fig. 1) Dr. Vittoria has also extensively researched comparable forms for the most diagnostic among the black gloss fragments (approximately 2000 in number), using them to derive a more detailed chronology for approximately 250 of the 610 total sites. Richard Fletcher (computer expert and doctoral candidate, University of Sydney, Australia) has been brought aboard to develop a database that will store essential information for all studied artifacts, beginning with the black gloss. (See "Dating the Results of the Metaponto Field Survey," pp.50–52.) Photographic documentation of the black gloss ceramics has been completed by professional archaeological photographer Dr. Giambattista Sassi (University of Lecce). Both the photographs and the illustrations have already been prepared in digital format, ready for publication.

Cesare D'Annibale, field leader of the first phase of the survey (1981–1990), has returned to study the prehistoric artifacts (lithics and pottery) and develop a chronology and typology of the prehistoric sites. He is updating his typology of the vast quantities of Greek coarse and table wares (jugs, bowls, pots, etc. meant for daily use) recovered during the surveys, with the purpose of identifying zones of local production that may help elucidate the nature of the rural ceramic industry. Marsha Robbins (doctoral candidate, Southampton University, England) is studying the Greek and Roman transport amphorae (see "The Greek and Roman Amphorae from the Metaponto Field Survey," pp. 65–66) while Smadar Gabrieli (doctoral candidate, University of Sydney, Australia) is preparing a section on the Greek and Roman kitchen/cooking wares for the publication of the Pantanello Sanctuary and the survey volume (see "The Kitchen/Cooking Wares from the Metaponto Field Survey," pp.59–62). Prof. Rebecca Ammerman (Colgate University, New York) and Mary Malone (UT Austin) are collaborating on a study of the terracotta figurines from both the survey and the Pantanello Sanctuary (see "Terracotta Figurines and Plaques from the Pantanello Sanctuary and Metaponto Field Survey," pp. 62–64). The last major categories of Greek artifacts, the architectural terracottas and Red-figure pottery, are being studied by Anna Lucia Tempesta (doctoral candidate, University of Naples). These last studies should reveal patterns of religious practices and house construction among the farmers of Metaponto.

For the Roman period, the Republican-era greyware ceramics have been studied by Dr. Cracolici, and the Imperial finewares—Arretine ware, African Red-

Slip, etc.—by Erminia Lapadula (doctoral candidate, University of Lecce), who has also finished the study of the Late Antique, Byzantine, and Medieval ceramics (see "Roman and Post-Roman ceramics from the Metaponto Field Survey," pp. 56–58).

Preparation of the publication itself is already at an advanced stage. Once the artifact studies have been completed and submitted (late spring 2003), analysis of the results will begin immediately so that the core sections on settlement patterns in the various periods (Prehistoric, Greek, Roman, and Post-Roman) can be written (Chapters 5–6 and 8–9). Meanwhile, Chapter 2, "Structural and Developmental Geology of the Metapontino" by UT geologist Robert L. Folk, will be based on Prof. Folk's report of his on-site researches conducted between 1980 and 1987; this chapter is currently in the editing stage. Chapter 3, "Geoarchaeology and Geomorphology of the Metapontino" by James T. Abbott (PhD, UT Geography, 1997), has been edited and stands ready for publication. The final section of the book, the Gazetteer of Sites, has been prepared by Alberto Prieto (doctoral candidate, UT Classics) and Gordon Lobay (ICA). The remaining chapters (such as Chapter 4, "Archaeological Survey Methods," and Chapter 7, "Greek Organization of the Physical Landscape at Metaponto") will be addressed during the course of the year by the relevant specialists. This will include a section of the famous grid of "division lines"—the first known instance in the Greek world of plots that later characterized other chorai, such as that of Chersonesos—as well as a discussion of the methods used to discover and map them. We are hopeful the first draft of the publication will be completed by the end of 2003.

Dating the Results of the Metaponto Field Survey

By Richard Fletcher
University of Sydney, Australia
Alberto Prieto
University of Texas at Austin/ICA
Ted Robinson
University of Sydney, Australia

The Artifact Database

As the foundation for data storage, retrieval, display, and manipulation, database software can by now be found in every archaeological project of any size or scope. Databases are essential for managing and analyzing the vast amounts of data generated by excavations and field surveys, usually describing architectural elements, archaeological strata, site characteristics, and artifacts. ICA has long employed databases in the Metaponto field survey: in the late 1980s separate databases were created for (1) the geographical and functional data associated with each site (e.g., geographic coordinates, distance from the ancient urban center, site type, etc.) and (2) the vast range of artifacts collected at each site. The former was re-evaluated in 1998-1999, revised, and then imported into the Metaponto GIS to serve as a "sites" layer for visual display, querying, and analysis. The intensification of artifact study in 2002, however, has rendered the second database obsolete due to the introduction of modifications to artifact classification and chronology in the intervening years, and because of the very fact that a new generation of specialists, with new interests and approaches, are studying the materials. The publication of the first survey volume has therefore provoked the creation of a new artifact database.

The new database is designed to assemble in one location the essential data for all artifacts of chronological or morphological significance, in whatever artifact classes they may fall, and thus it will serve as the primary tool for analyzing the geographic and chronological distributions of individual forms and groups of related forms. The patterns detected in the ceramic distributions will in turn shed light on the functional and chronological assignments of the sites themselves. To maintain compatibility with most GIS software (a crucial consideration now for every archaeological project), the database has been created in the industry-standard Microsoft Access software. All data is contained within one central table, in which each artifact occupies one row (or record). (Fig. 1) The data for

ArtefactID	Site ID	Find Date	Box	Mat	Classification	Form	Part	Type
1	1		140	FS.81.1PL.k	Red Figure	Krater	Foot	
2	1		140	FS.81.1PL.a	Banded Ware	Pl	PR	
3	1		140		Banded Ware	Amphora	Rim	
4	1		140	FS.81.1PL.c	Black Glaze	Skyphos	Foot and body	Attic
5	1		140	FS.81.1PL.f	Black Glaze	Skyphos	Rim	Corinthian
6	1		140	1	Black Glaze	Cup	Rim	Ionic B2
7	1		140	FS.81.1PL.e	Black Glaze	PY?LG?	RSh	
8	1		140	2	Black Glaze	Cup	Rim	Ionic B2
9	1		140	3	Black Glaze	Skyphos	Rim	Corinthian
10	1		140	4	Black Glaze	K	Rim	Bloesch C
11	1		140	5	Black Glaze	K	Rim	Bloesch C
12	1		140	6	Black Glaze	Skyphos	Rim	Attic
13	2		140	FS.81.5PL.p	Black Glaze	Ld	knob	
14	2		140		Black Glaze	Cup	Rim	Ionic
15	4		140	FS.81.4PL.b	Black Glaze	Patera	Foot and body	
16	4		140		Black Glaze	C-sk?	Foot and body	
17	4		140	FS81.4PL.f	Black Glaze	Skyphos	Rim	Corinthian
18	4		140		Black Glaze	Patera	Base	

Fig. 1: A portion of the main data table.

each artifact (residing in the columns) is entered via a customized visual portal specific to each class of artifact; there are portals for finewares, coarsewares, kitchen/cooking wares, transport amphorae, lithics, and other materials. Within each of these portals information about the artifact is visually arranged in distinct sections based on logical association (e.g., a section for all information pertaining to physical description, a section for chronological information, etc.). (Fig. 2) Different classes of artifact require different information: for example, while the finewares portal has fields (or data containers) for only *Site*, *Artifact*, *Date* and *Notes*, the coarsewares portal has in addition to these a field for *Fabric*, and the lithics portal has a section with fields tailored to the description of stone tools.

In order to allow a GIS to match the geographic information for each site (in the "sites" layer) with the associated artifacts (in the artifact database), the site and artifact data share a *Site Number* field. The columnar fields containing artifact information will allow queries to be performed in the GIS on the basis of a number of different criteria. For example, under finewares, the *Category* field will permit searches on the basis of "Drinking Vessels" or "Pouring Vessels," while the *Sub-category* field will allow for a search on the basis of the particular shape of drinking vessel (e.g., cup-skyphos) or the particular type of pouring vessel (e.g., oinochoe). The same function applies to all the other fields in the *Artifact*, *Fabric*, and *Lithics* sections. Each row/record also includes a field for linking digital photographs and illustrations of the artifact itself or a comparable artifact.

The "Date" section contains fields for *Probable Date*, *Date Range*, and *Chronology*. In most cases, only the finewares will have reasonably accurate dates. The *Probable Date* field is not intended as a strict indication of date; it is used for rapid and convenient illustration of the data in the GIS and to facilitate queries within the GIS. Thus, a date-range of 400-375 BC may have a "probable date" of 380 so that the GIS can readily assign the artifact to this quarter-century period. Any date range greater than 30 years is left without an assigned "probable date" unless there are compelling reasons for overriding this convention. Date ranges greater than 30 years are stored in the *Date Range* field used for further analysis and/or display in the GIS. Finally, the artifacts that can be only very loosely dated make use of the *Chronology* field, which assigns a correspondingly coarse chronological classification such as "Archaic," "Classical," or "Hellenistic."

Dating the Material

The basis for dating almost all of the sites and materials will be the figured and black gloss pottery. The studies of Vincenzo Cracolici and his team will therefore serve as the very foundation for all chronological judgment and interpretation. Material which cannot be securely dated will be given approximate dates only if associated with dateable black gloss or figured pottery. However, the term "dateable" must remain rather loose. It is well known that archaeological dating almost always leaves a margin of error of at least ±25 years. Moreover, much of the materials from the survey, even the black gloss pottery of the 6th–3rd centuries BC, can often be dated to only a 50-year period. But depicting the changes over time at a given site requires something less than 50-year chronological units, since units so large may miss the more subtle changes between generations of settlers. Thus a 25-year division was chosen.

It was found, however, that little more than 9% of even the black gloss pottery could be dated to within the customary divisions of 1st, 2nd, 3rd, and 4th quarters

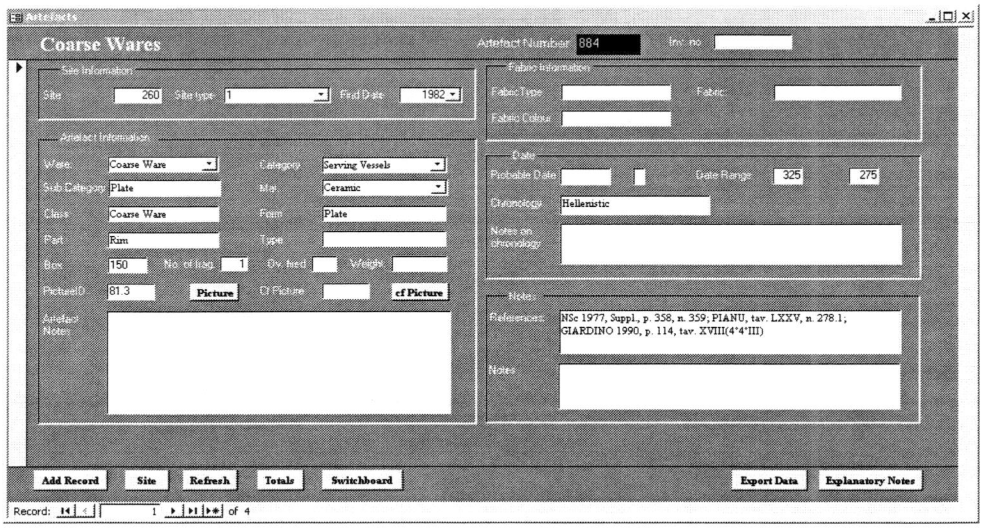

Fig. 2: The coarsewares data-entry portal.

of each century. Consequently a three-tiered approach was adopted, in which 50-year periods are displayed as overlapping units, 100-year periods are shown similarly overlapping, and intentionally broad chronological divisions such as "Archaic," "Classical," "Hellenistic," and "Roman" are also displayed (all dates BC) Such units can be added as long as there is chronological information, and they will allow a realistic depiction of change over time under the constraints of the Metapontine chronological sequence

<table>
<tr><td>750-700</td><td rowspan="2">750-650</td><td rowspan="10">Archaic</td></tr>
<tr><td>725-675</td></tr>
<tr><td>700-650</td><td rowspan="2">700-600</td></tr>
<tr><td>675-625</td></tr>
<tr><td>650-600</td><td rowspan="2">650-550</td></tr>
<tr><td>625-575</td></tr>
<tr><td>600-550</td><td rowspan="2">600-500</td></tr>
<tr><td>575-525</td></tr>
<tr><td>550-500</td><td rowspan="2">550-450</td></tr>
<tr><td>525-475</td></tr>
<tr><td>500-450</td><td rowspan="2">500-400</td><td rowspan="6">Classical</td></tr>
<tr><td>475-425</td></tr>
<tr><td>450-400</td><td rowspan="2">450-350</td></tr>
<tr><td>425-375</td></tr>
<tr><td>400-350</td><td rowspan="2">400-300</td></tr>
<tr><td>375-325</td></tr>
<tr><td>350-300</td><td rowspan="2">350-250</td><td rowspan="6">Hellenistic</td></tr>
<tr><td>325-275</td></tr>
<tr><td>300-250</td><td rowspan="2">300-200</td></tr>
<tr><td>275-225</td></tr>
<tr><td>250-200</td><td rowspan="2">250-150</td></tr>
<tr><td>225-175</td></tr>
</table>

Black Gloss Pottery from the Metaponto Field Survey

Vincenzo Cracolici, PhD
Independent Archaeologist
Bari, Italy

The Study: Aims and Methods
The majority of artifacts collected during the intensive field survey of the Metapontino consists, naturally, of pottery, and among the various classes of pottery, arguably the most important is the black gloss vessels. Although the chronological definition of certain shapes within certain periods remains problematic, black gloss pottery nevertheless serves as the best criterion for dating archaeological sites. Detailed study of this pottery is therefore the key to determining settlement patterns in the colonial Greek period.

The first step in the study was the assignment of a general chronological range to each site based on a general impression of its diagnostic black gloss fragments. (Sites containing Roman "greyware" pottery, which can be only roughly assigned to the 2nd–1st centuries BC, are dated by simple presence/absence of that material.) This initial analysis generated an overview of the entire corpus of artifacts and identified sites containing material of relatively greater importance, which would become the subjects of more focused study. Thus the second step —recently completed—was a close analysis, for even greater chronological and typological precision, of approximately 2000 diagnostic black gloss fragments from 250 selected sites. The data derived from this second step may lead to revisions of the general chronology.

The ultimate goal of this study is a typology with well-defined absolute chronology, creating a definitive picture of the local ceramic industry throughout the Metapontino. The system employed is an open one, to which new types can be added in the future as they are identified. The black gloss analysis will serve as the overarching framework into which the detailed studies of the remaining ceramic classes (undecorated dining wares, storage and transport vessels, kitchen/cooking vessels, etc.) will be integrated.

Shapes and Chronology
Artifact assemblages of the Archaic period (600–490 BC) are generally characterized by an absence of distinctive shapes. The majority of the recovered ceramics appears to be locally-produced, with very few imported items. The best diagnostic shape for the entire 6th century BC is the "Ionic Cup" (Fig. 1). Later in the century the most common shape is the type B2

cup, present in many variants representing divergent traditions among both workshops and individual artisans. No other characteristic shapes of the first half of the century, such as the type A or B1 cups, have been so far identified. A final form worthy of note is the "Ionic" or "Pantanello" skyphos, with an offset rim and painted decoration closely related in style to those of the B2 cups.

In the Classical period (490-323 BC) we can begin to trace the evolution of certain shapes, and likewise the introduction of shapes unknown in the Archaic period permits greater chronological precision. In this period the "Pantanello" skyphos becomes more rigid in shape; the rim is sharply distinguished from the body, and the gloss becomes shinier and thicker. Forms such as the olpe (a type of jug) and stemless cups appear for the first time. (Fig. 2) Two new forms of skyphos, the "Attic" and "Corinthian," are introduced and immediately begin to evolve. In many forms the base mouldings mutate more rapidly than the rims, so that the bases serve as a more reliable source of chronological information.

For the end of the 5th century BC onward we are fortunate in having a considerable quantity of comparable material from the *kerameikos* (potters' quarter) of Metaponto. The repertoires of the various urban workshops expand noticeably, and there are reflections of this change in the survey assemblages. Many shapes originating earlier in the 5th century develop into easily-recognizable types with clear evolutionary histories. Ribbed and rib-less concave dishes, Red-figure vessels, and small bowls, for example, found in great quantities in the *kerameikos*, also appear throughout the chora and suggest a flourishing extra-urban production. (Fig. 3) The evidence from the survey demonstrates unequivocally that the ceramic production of the chora was as rich and varied as that of the urban center itself.

In the second half of the 4th century BC, again relying heavily on evidence from the *kerameikos*, we observe a standardization of the mode of production whereby the internal shapes of vessel-bases change as a result of the rapid diffusion of kiln firing-supports, a feature that had been introduced at the end of the 5th century. Close analysis of the *kerameikos* artifacts has demonstrated that the supports were initially designed to accommodate existing types of bases, so that the range of support-shapes was dependent on the range of vessel-shapes. Beginning in the middle of the 4th century, however, a reverse process takes effect, simplifying the relationship between vessel and support: vessel-bases are designed for use with the firing-supports, reducing the number of support types from approximately 50 at the beginning of the century to just three by the beginning of the following century. (See V. Cracolici, *I sostegni di fornace dal kerameikos di Metaponto.*

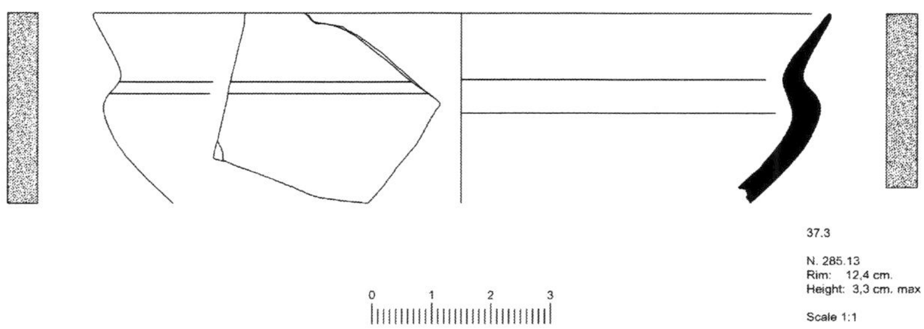

Figure 1: Rim of a black gloss "Ionic" cup. 600–550 BC.

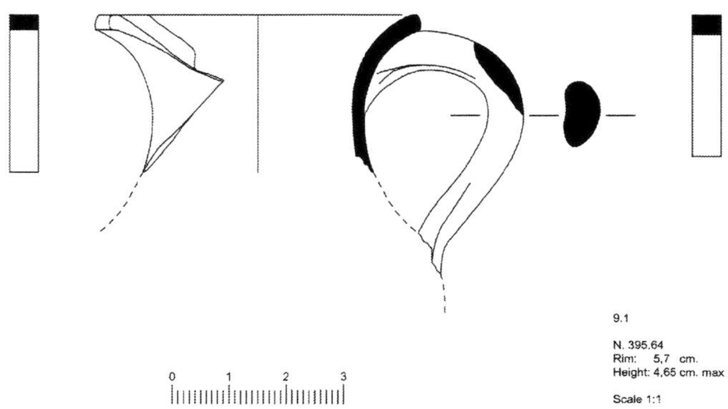

Figure 2: Rim and handle of a banded ware olpe. 310–280 BC.

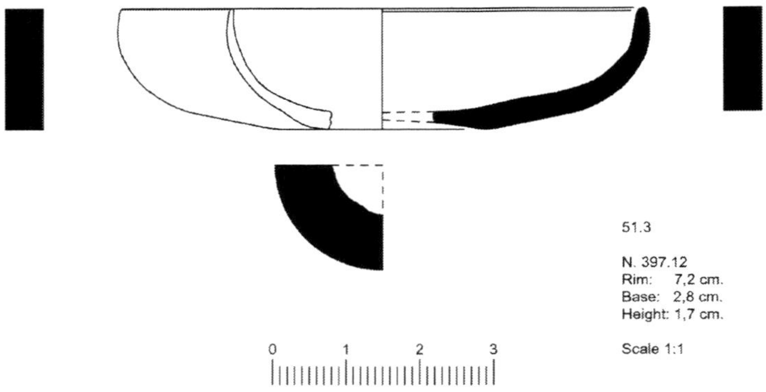

Figure 3: Small black gloss bowl. 310–250 BC.

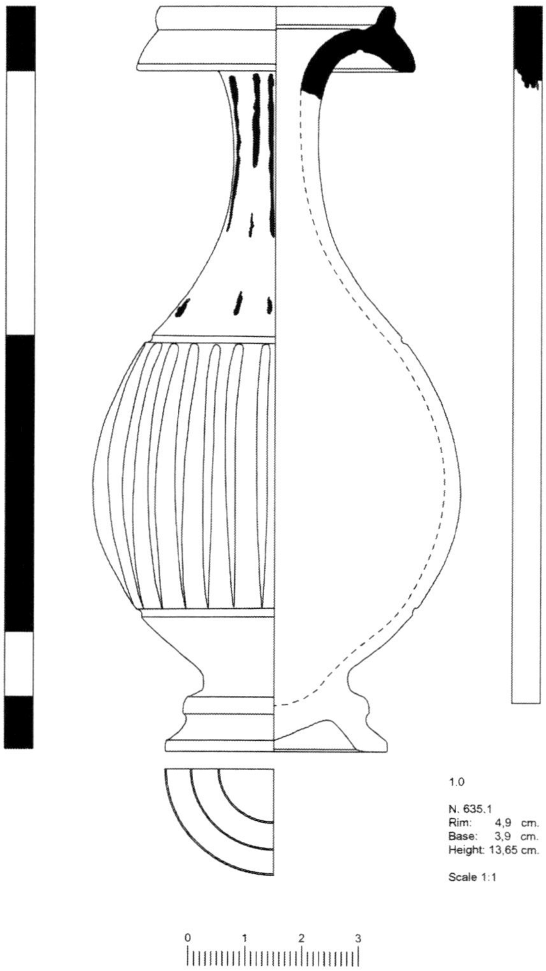

Figure 4: Gnathia-style unguentarium. 325–280 BC.

Aspetti tecnici e modi di produzione della ceramica dall'età arcaica all'ellenismo [PhD Dissertation, University of Naples, 1998] and "Le ceramiche," in A. De Siena, ed., *Metaponto. Archeologia di una colonia greca* [Taranto 2001], 103–114.) Therefore, though bases are of greater use in dating earlier materials, by the second half of the 4th century the rims (and forms in general) serve as better chronological markers.

The Hellenistic period (323–100 BC) enjoys the richest variety of open shapes, providing a good basis for typological classification. There are considerable problems of chronology, however, since this period is rather poorly understood, and ceramic scholarship has made little effort to distinguish shapes of the 4th and 3rd centuries BC. There is a widespread tendency among archaeologists to assign material to a general 4th/3rd century category, and the problem is often extended into successive periods so that 2nd century BC material, with a few notable exceptions, is all but unidentifiable. Casualties of this unfortunate situation are, for example, unguentaria, lamps, and lekythoi. (Figs. 4–5) Nevertheless, we can follow the final evolution of the "Corinthian" skyphos in the 3rd century: the base becomes a thick disk, while the rim curves out. (Fig. 6) Furthermore, the standardization of bases is complete and involves both open and closed shapes, so that vessels with preserved bases can be dated to the 3rd century, but no more precisely than this. Finally, a new shape appears, a plate with out-turned rim, the evolution of which has not yet been properly described. There are no traces of this shape in the *kerameikos* of Metaponto through the last quarter of the 4th century, when all production there ceases (though this is an *argumentum ex silentio*), but comparable fragments—with somewhat extended rims—have been recovered from excavations around Metaponto and dated to the 2nd–1st centuries BC. The fragments from the survey are therefore probably 3rd century examples; their rims may help define the evolution of this latest shape.

Ceramic Fabric

The color of the clay employed in all Metapontine ceramic production varies from pinkish-yellow to brown, running through a spectrum of reddish and beige shades. In black gloss vessels the fabric is usually very fine, with few inclusions, these consisting mostly of white dots—flakes of calcium carbonate-based rocks and slivers of gold or silver mica. (A precise qualitative/quantitative analysis based on a statistically valid sampling is forthcoming.) The texture of the fabric is at times smooth, at other times powdery, and occasionally hard. This variation is probably a consequence of the temperature and atmospheric conditions

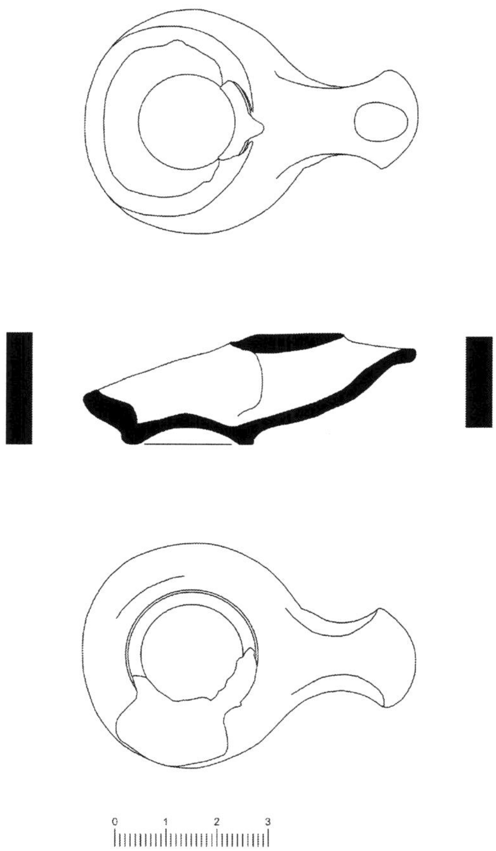

Figure 5: Black gloss lamp. 325–150 BC.

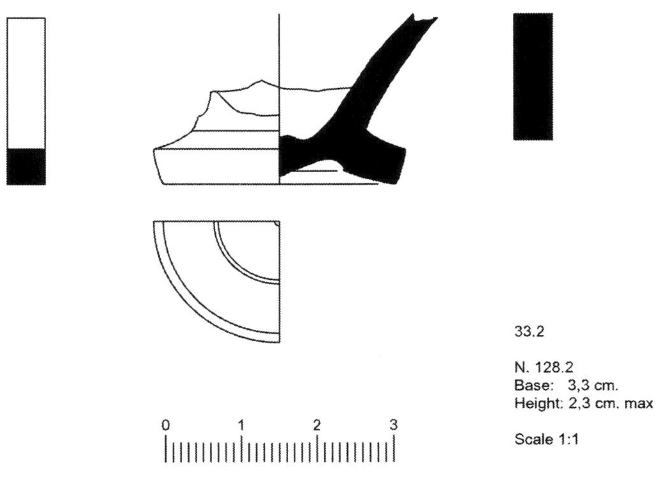

Figure 6: Base of a black gloss "Corinthian" skyphos. 320–280 BC.

inside the kiln during the firing process. The gloss itself is normally not very shiny, but virtually never matte. The color of the gloss sometimes verges into brown, most likely the effect of an imperfect reduction (oxygen-deprivation) phase during firing. An important, but unfortunately not definitive, distinction can be made on diachronic grounds: Archaic vessels are very often powdery to the touch, with matte gloss, while vessels of the Classical period are usually hard and covered by a thick film of shiny, metallic gloss.

It is impossible to distinguish fabrics produced in the urban center from those produced in the chora, despite excavation of the *kerameikos* of ancient Metaponto and numerous other sites in the chora (some of which, especially farmhouses, can be considered possible production sites unto themselves). As far as one can judge, the examples from the chora are identical in both shape and quality to those produced in the *kerameikos.* Thus it is clear that the urban workshops alone could not have been responsible for the chronological and typological range of vessels found in both the chora and the urban center. The general impression gained from both urban and rural necropoleis is of a possible distinction between high- and low-quality wares, rather than a real difference in fabrics. Therefore considerations of "urban" and "non-urban" fabrics cannot be justified.

At the risk of making the argument appear circular, it is possible to argue that the difficulties in distinguishing urban and rural products are the result of a general tradition of craftsmanship common to the entire area and strongly influenced by the urban center itself. By ca. 350 BC the process of production is uniform across the chora and even outside of it, and black gloss vessels are produced in all areas until the Roman conquest of southern Italy. There are two possible explanations for these apparent similarities. The first is a certain uniformity in the local clays, an idea rendered quite plausible by the homogeneous character of the local geological structures and the great abundance of suitable raw material. This option, however, fails to explain the wide divergence between indigenous and colonial Greek ceramic production in southeastern Italy. Thus we must turn to the second possibility, tangential to the first: a common method of purification, treatment, and firing of clays. This, too, must be viewed as part of a tradition of pottery craftsmanship.

Roman and Post-Roman Ceramics from the Metaponto Field Survey

By Erminia Lapadula
University of Lecce, Italy
[Translated by Alberto Prieto]

Introduction

The ceramic artifacts belonging to vessels of the Roman and post-Roman periods are very limited in number when compared to the enormous quantities of material collected by the survey crews on the surface, most of which pertains to the Greek period. Moreover, the majority of these artifacts are of very small size and often belong to morphologically unimportant or otherwise non-indicative parts of the vessel. Nevertheless, beyond indicating the presence/absence of the ceramic classes, the study of these materials will allow us to propose various interpretations of the Metapontine landscape in the Roman and post-Roman eras.

Roman Ceramics

The sites of the Augustan and Early Imperial eras (approximately 50 BC–AD 50) are only 12 in number. The most important class of ceramic materials among them is the fine table wares, represented by the *Terra Sigillata Italica* and Eastern Red *Terra Sigillata* wares; these are accompanied by a few fragments of undecorated table wares and amphorae. Noteworthy among the fragments of *Terra Sigillata Italica* is a cup-base bearing the stamp "AUC[TUS]" inside a rectangular border, found at Site 428. (Fig. 1)

Equally scarce are the ceramic materials of the high Imperial era (second half of the 1st century AD – 3rd century AD), the majority of which are preserved in fragments of modest size. The best-represented classes are those of African origin, both fine table wares and kitchen/cooking wares, with fairly consistent proportions. The most common forms are generically datable to the period ca. AD 150-250. The collections of fragments of Eastern Red *Terra Sigillata* wares, undecorated table wares, kitchen/cooking wares, and transport amphorae are very limited. This period is represented by 25 sites.

The evidence pertaining to settlements of the Late Antique period (4th century – first half of the 6th century AD) also encompasses a total of 25 sites. The most prominent of these sites, for the quantity and variety of ceramics attested, is Site 230, which continues without interruption from the Early Imperial period. Site 230 also contains ceramics of both local manufacture (kitchen/cooking wares, decorated table wares) (Figs. 2 and 3) and import (African and Phocaean *sigillata* wares, African and Eastern Mediterranean transport amphorae), in equal proportions.

Among the Late Antique ceramic classes attested, particular importance is assigned to the decorated table wares (otherwise known as "Calle"-style ceramics, from the production site discovered in the vicinity of Tricarico, Province of Matera). Designed for table and pantry, these wares are indicators of a lively local (regional or sub-regional) production. At the sites identified in the Metapontino, painted ceramics are present at rather high levels, although not as high as those of imported vessels. The range of shapes is limited, with a slight predominance of the open forms (basins, bowls, cups) over the closed (amphorae, jugs).

Figure 1. Base of Terra Sigillata Italica (Arretine) ware cup bearing stamp "AUC[TUS]." Site 428. Late 1st c. BC–early 1st c. AD. (Photographs courtesy of the author.)

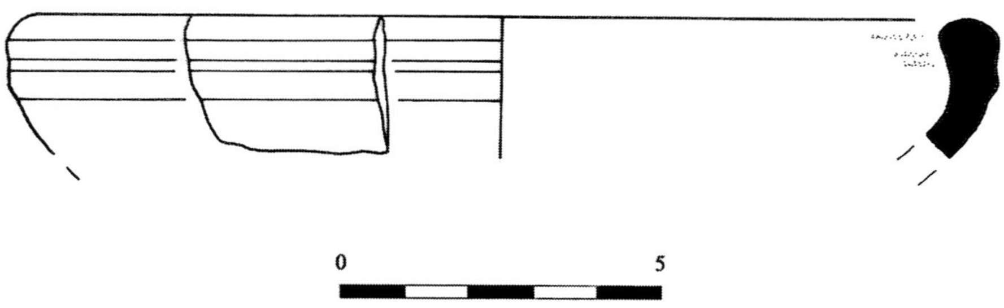

Figure. 2. Painted coarseware bowl. Site 230. Late 4th–5th c. AD.

Medieval Ceramics

The comparatively few sites of the Early Middle Ages (three) contain a very small number of fragments of undecorated table wares (cauldrons, jugs) and banded wares. Nevertheless, the presence of fragments of closed vessels (amphorae) with wide painted bands (Fig. 4) suggests a date for these settlements towards the final centuries (10th-11th) of the Early Middle Ages. The sites occupied in the Late Middle Ages are numerous (30 sites). They are documented predominantly by undecorated table wares (59.26%), followed distantly by kitchen/cooking wares (14.20%). Lead-glazed table wares are attested (11.35%) by polychrome glazed ceramics in three colors (dark brown, green, and red) or in varying combinations of colors (dark brown and green, dark brown and red, dark brown alone), by monochrome green ceramics (11.97%), and to an almost invisible extent by monochrome ceramics in light brown (0.90%) or yellow (0.34%). "Proto-Majolica" wares are limited to just two fragments of bowls (0.14%), and similarly low proportions are observed in incised polychrome wares (1.88%), narrow-banded wares (0.41%), double-dipped wares (0.55%), and transport amphorae (0.41%).

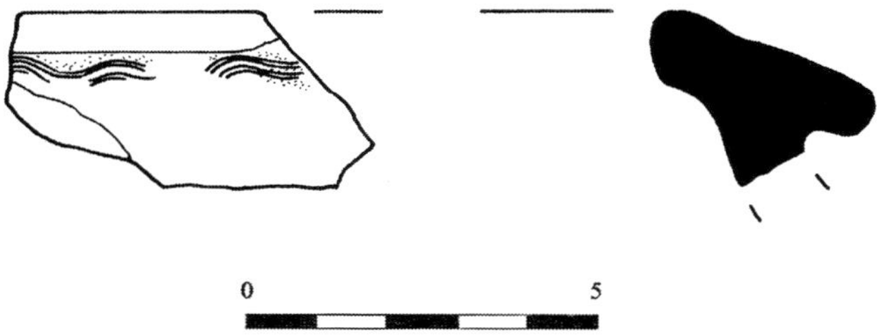

Figure 3. Rim of painted coarseware basin. Site 230. Late 4th–5th c. AD.

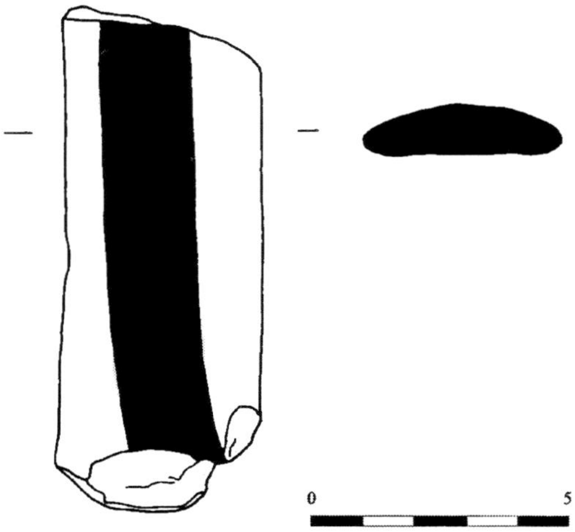

Figure 4. Handle of table-amphora with wide painted band. Site 326. 9th–10th c. AD.

The range of Late Medieval ceramic forms attested by the survey assemblages differs little from the typical range of Late Medieval ceramic classes encountered in southern Italy generally. Among the vessels designed for kitchen use only two forms are documented, the pot (in various sizes) and the pan. The undecorated pantry and table wares are represented by a large number of closed containers (amphorae, jugs, and bottles), along with open forms (bowls, mugs, basins). Noteworthy is a large number of pantry amphorae, often equipped with lids; these are peculiar in having two ribbon-shaped handles of consistent thickness and a flat bottom, demonstrating that they were intended to contain water, although storage of solid foods such as grains cannot be excluded as a function. The form is widely attested in the Salento (the "heel" of Italy) and throughout southern Italy generally. In the Metapontino, however, it never appears with the painted decoration common on the products of neighboring areas. Also noteworthy among the undecorated wares is the presence of several closed vessels equipped with a rounded basket-handle. The form is typical of Early Medieval cauldrons in kitchen/cooking fabric, but the function of the Metapontine vessels, which are characterized by a carefully levigated fabric precluding kitchen use, was more probably the transport of water rather than the preparation of foods or the warming of liquids near a source of heat.

Table wares with water-proofing treatments (monochrome and polychrome glazes, "proto-Majolica," double-dipping) are present primarily in open forms (bowls and, in lesser quantities, mugs and basins), while closed forms (bottles and jugs) are documented by few examples. The polychrome glazed ceramics display decorative patterns with relatively close parallels among the materials recovered in numerous settlements of southern Italy, in Puglia (Lucera, Taranto, Otranto, Brindisi, environs of Bari, environs south of

Figure 5. Base of polychrome glazed bowl with "Taranto" style decoration. Site 300.

Lecce), Basilicata (Torre di Mare, Policoro, Monte d'Irsi, Satriano, Montescaglioso), Calabria (Scribla), and the general Mediterranean area (Greece, Albania, Croatia, Malta, Israel). Among the most widely-attested decorative motifs are the "Taranto" style (Fig. 5) and the "Torre di Mare" style (Fig. 6), characterized respectively by a central medallion divided into two sectors populated by wavy lines in the tondo, and by a motif of concentric circles enclosing a wavy line on the interior wall beneath the rim. Equally diffuse is the "gridiron" motif, found on both products with lead-based decorative treatment and an example of "proto-Majolica." (Fig. 7) There are also fragments of vessels decorated with animal, vegetal, and geometric motifs.

The fabrics utilized for the production of both decorated and undecorated wares are substantially the same. The clays are well-levigated, dense and hard, and light in color (varying from pale yellow to beige). The rare inclusions are small and primarily white, occasionally red or dark, and frequently include mica; they probably originated in neighboring areas connected to local (regional and sub-regional) manufacturing activities. Production centers for both monochrome and polychrome glazed ceramics are known throughout Puglia (Apulia) and, to a lesser extent, in Campania (not to mention the more well-known Sicilian workshops). But it is to the Apulian manufacturing traditions, and in particular those of Taranto and the Salento, that the Metapontine "market" probably looked for inspiration. (To the same traditions, perhaps, should be ascribed the only two recovered fragments of "proto-Majolica," parts of two bowls from Site 300). (Fig. 7)

Ceramics imported from places outside of southern Italy in the Medieval period are represented by a single artifact of incised polychrome ware, a bowl-rim bearing an incised motif of spirals on the interior wall beneath the rim itself. The fabric is distinguished from that typical of the region by its color (red) and the relatively high density of its dark calcite and mica inclusions. The interior surface is covered with a cream-colored slip, while a green and brown scheme decorates the rim and the area beneath the incised motif. (Fig. 8) This artifact possibly originated in the upper Adriatic region.

Post-Medieval Ceramics

Post-Medieval ceramics are present in the survey assemblages at extremely low percentages: overall approximately 1.03% of the total number of *Medieval* artifacts recovered by the survey. Only four sites contain significant quantities of ceramic material dated later than the 16th century, although isolated fragments of post-Medieval wares are found in numerous Early Medieval sites, attesting the continued frequentation of some sites into the Modern period. The ceramic classes characteristic of the post-Medieval sites are primarily undecorated table and kitchen/cooking wares (both decorated and undecorated), along with a smaller number of fragments of incised glazed wares. The forms of undecorated table wares are open and closed, of small and large dimensions; the kitchen/cooking wares are attested exclusively by pans and pots. The incised glazed wares, however, are characterized by plates and bowls.

Figure 7. Bases of bowls with "proto-Majolica" decoration. Site 300. Mid-3th c. AD.

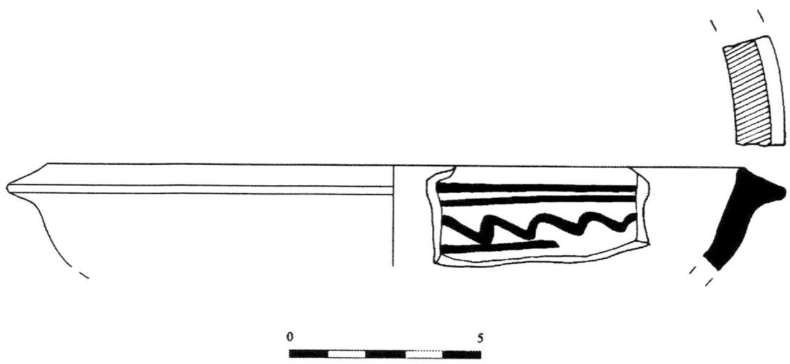

Figure 6. Polychrome glazed bowl with "Torre di Mare" style decoration. Site 33. Late 13th c. AD.

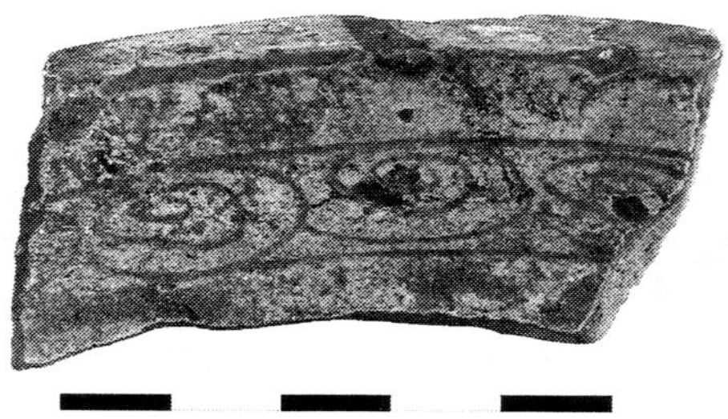

Figure 8. Bowl with incised polychrome decoration. Site 381. 12th–13th c. AD.

The Kitchen/Cooking Wares from the Metaponto Field Survey

By Smadar Gabrieli
University of Sydney, Australia

Introduction

In the summer of 2002 Professor Carter suggested that I extend my study of the kitchen wares from the Roman kiln deposit at Pantanello to include the kitchen wares recovered by the field survey of the chora. This opportunity to look at material from a completely different context required rethinking the research approach. In the 2001 Annual Report, I discussed the study of the kitchen wares from the kiln deposit, which accumulated over a period of approximately 150 years. Such extended, stratified accumulation is necessary for a detailed study of change and continuity in manufacturing technology, production modes, and patterns of cooking habits. Survey material, on the other hand, though it has a similar or even identical range of shapes, requires formulation of a completely different set of questions to overcome the absence of chronological sequence and firm association between various categories of material culture.

Excavations and surveys are complementary: excavation of a single site provides a detailed profile, albeit in isolation ("micro-scale"); survey provides the context of interaction between the various sites within their surrounding landscape, but it lacks detail ("macro-scale"). In an excavation it is possible to quantify the various types of kitchen wares in successive stages in order to understand the changing (or consistent) patterns of their use: a shift in preference from stews to roasting, for example, will result in a parallel shift from deep to shallow cooking vessels. In surface collections, where the sample is more random, and where stratification is entirely lacking, quantification is of limited use. Nevertheless survey collections provide data on presence and/or absence of specific types of material culture over a wide geographic extent and are therefore essential for interpreting excavated sites. The character of a single site can be fully understood only when considered in the context of its surroundings.

As I stated earlier, the leap from a single site to survey material requires a mental shift. The main aim of this project so far has been to formulate questions which, once integrated with the rest of the evidence, will help us understand the nature of ancient use of the landscape. This section will therefore present not so much a report of the results of a study, but rather an outline of research in progress, an explanation of an approach. Questions will be posed, with only a hint of answers.

Kitchen wares as an interpretative tool

Information from kitchen wares collected in survey can be used to interpret the functional orientation of not only the individual sites but also the territory as a whole, with the constant caveat that kitchen wares, like any class of artifact, must be studied as part of the overall archaeological assemblage, not in isolation.

On the level of individual sites, kitchen wares contribute to the identification of the character of a given site: the composition of wares in a sanctuary, for example, should be different from that of a household. The kitchen wares of a seasonal occupation will differ from those of a long-occupied farmhouse. Sites with mixed use will be different again. The presence or absence of kitchen wares, and their quantitative relations to other types of wares, aid in interpreting the use(s) of a site.

The potential information inherent in kitchen wares becomes particularly interesting, however, when the focus shifts to the territory as a whole. One of the most intensively-debated issues in Western Greek archaeological research is acculturation, i.e., the rate and extent of the assimilation of a new culture, upon both the arrival of the Greek colonists and the subjugation of the area by the Romans. Acculturation begs questions such as: to what extent did the newcomers supplant the existing culture? and, on the other hand, how were the newcomers influenced by the existing culture? In other words, how does one assess the balance of forces among superimposition of a new culture over an existing one, integration of the new population into the local one, and acculturation of the existing population to the new?

Food preparation, and the use of specific kitchen utensils, is widely considered a particularly conservative aspect of pre-modern societies. Practically everyone wants to keep up with the Joneses in matters relating to conspicuous consumption, but when it comes to food, "mother knows best." An engaging example of this attitude is found in a documentary film on traditional pottery production in Cyprus by Gloria London (*Women Potters of Cyprus* [VHS; Nicosia, 2000]). In an interview with a newlywed couple buying a pot from a specific potter, the husband explains that his mother has sent them: if he want his new wife to cook like her, she said, he should start by purchasing the exact same pot used at home. This attitude is useful when one looks at acculturation.

When food preparation undergoes a significant change, reflected in the introduction of new ceramic forms for the kitchen or pantry, there is no doubt that acculturation has reached a fundamental stratum of society, while the introduction of, for example, symposium vessels may signify only the arrival of a new fashion.

The study of the kitchen wares from the field survey is dependent on eventual integration with the other aspects of study for a meaningful interpretation – indeed, much more so than the study of the kitchen wares from the ceramic production site at Pantanello. The potential results are exciting. We expect that, in combination with the remaining evidence, the kitchen wares will help define the changing nature of the chora and the relation of the changes to location within the landscape (for example, proximity to the urban center or major roads).

Preliminary results

When approaching kitchen wares using quantitative methods to provide insight into ancient eating habits, three minimal assumptions must be accepted:

(1) All types of kitchen wares (bowls, pots, pans, casseroles, and jugs or jars made of cooking fabric) have similar rates of physical preservation and archaeological recovery. These vessels generally have the same size, wall thickness, and fabric – and therefore similar breakage patterns – with consequently comparable preservation and recovery rates.

(2) Given (1), and as long as the collection methodology remains consistent, the relative frequencies of open and closed vessels can be successfully compared across sites.

(3) Storage wares and large food-processing installations—e.g. wash basins, large mortars, and pithoi—are considerably more robust than average kitchen wares and thus have better preservation rates. They are also more conspicuous on the surface and will therefore have higher retrieval rates than kitchen wares. Consequently, their presence or absence in conjunction with kitchen wares is related to the use of space.

Both (2) and (3) are valid only as long as vessels can be assigned with certainty to specific periods (always a difficulty with kitchen wares), or if trends that remain constant over long periods can be identified.

To date, the study has addressed the materials from 377 sites spread over the entire survey area. Only 96 of these sites contain identifiable fragments of kitchen wares, but a much larger proportion of the sites have fragments of storage and large food-processing vessels, supporting assumption (3) above. In the sites containing kitchen wares, representation of the pre-Greek, as well as the Late Antique and later periods is nearly non-existent, so that the following remarks and the illustrations of representative shapes, pertain only to the colonial Greek period (6th–3rd centuries BC). Of the abovementioned 96 sites, only 26 (27%) contain ten or more identifiable sherds of kitchen vessels. All of these sites are relatively large (more than one hundred sherds collected), but not all large sites contain kitchen wares.

Considering only the cooking vessels, casseroles are the dominant shape in most sites, and they demonstrate by far the most variation in their shapes. In 20 out of the 26 large sites, however, the dominant shape is pots/jars. In fact, the impact of these 20 sites is such that, in an overall count of cooking vessels, pots/jars are the majority.

Before addressing the significance of these findings, we should consider the range of kitchen vessels identified so far. The open shapes include, first and foremost, casseroles (Fig. 1) – shallow, covered vessels used for roasting, simmering, and cooking large, flat portions of meat, such as whole fish. The constraints of cooking in casseroles are rather obvious from their shape: the height is rarely more than 5cm, the walls nearly vertical, so that the rim is as wide as the base, while the lid is supported on a small shelf. The Metapontine casseroles have two small handles hugging the walls, suitable for providing a grip while removing the vessel from the fire, but not for easy and rapid manipulation over the fire (as is required, for example, when frying). The flat bases may indicate that they were used on a flat cooking surface. A variant, with loop handles rising above the rim – for which we have no complete profile, but nevertheless a clear indication of a round base (Fig. 2) – is closer in shape to the classic Greek *lopas*. It may either belong to an early stage of that vessel or represent a somewhat different mode of cooking.

Other open shapes include pans (Fig. 3); flat dishes (Fig. 4), which may or may not have been used for cooking; bowls (Fig. 5), which are fairly rare; and mortars, sometimes equipped with a spout (Fig. 6). The frequent appearance of mortars may help to explain the paucity of bowls in kitchen ware fabrics: mortars were not used exclusively for grinding and chopping. Cato's instructions in his recipe for cheese and honey tart (*De agri cultura* 76) include draining the cheese and kneading the dough in a mortar, obviously assuming its use as a bowl.

The closed shapes are represented by pots/jars (the terms are often used interchangeably) (Fig. 7), used to boil water or cook liquid and semi-liquid foods, such as stews or cereal porridges, and jugs (Fig. 8). All three shapes may be used either as storage or as cooking vessels, but normally the pots with rounded bases (e.g. Greek *chytra*) are associated mostly with cooking, and the jugs with storage of liquids, while the jars (e.g. the Roman *olla*) are truly bi-functional.

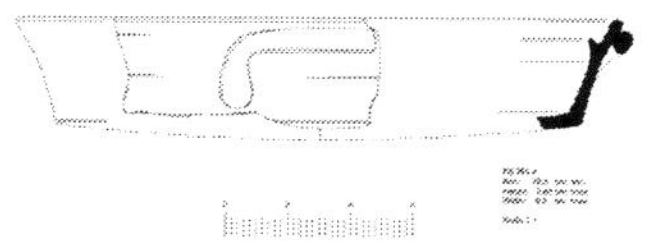

Figure 1. Casserole.

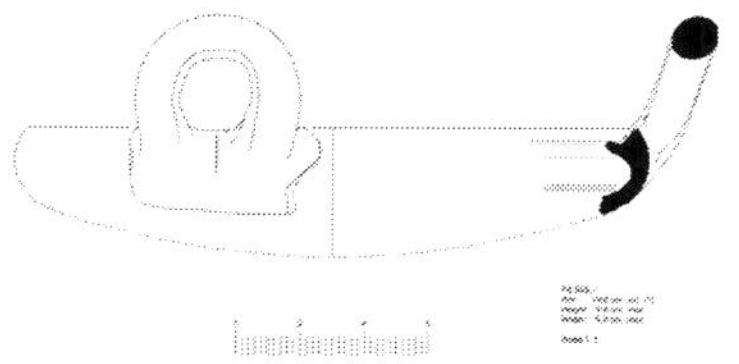

Figure 2. Casserole with loop handle.

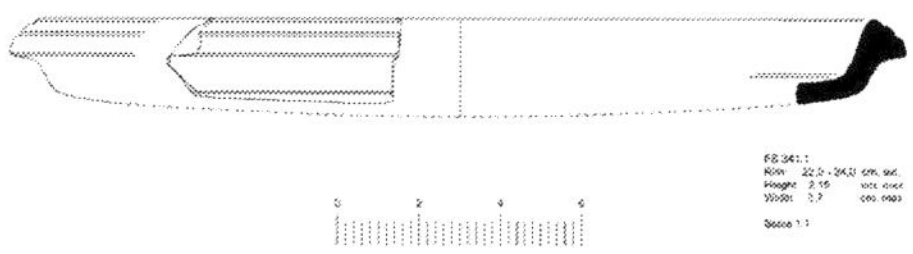

Figure 3. Pan.

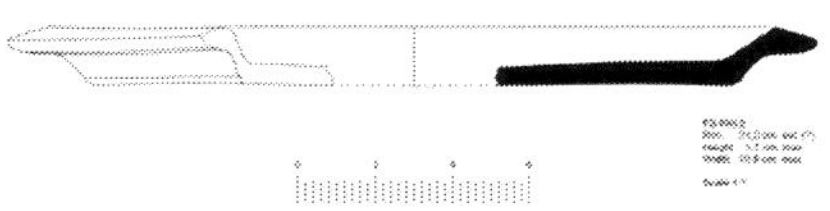

Figure 4. Dish/plate.

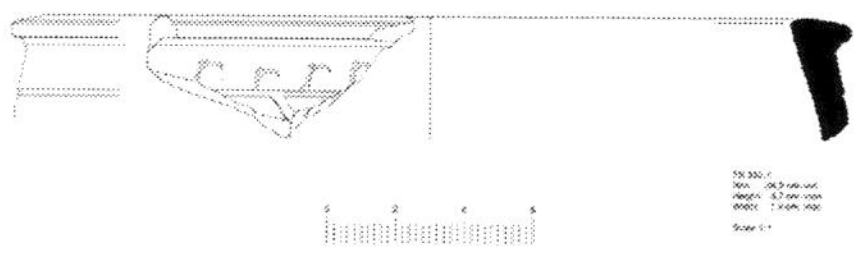

Figure 5. Bowl.

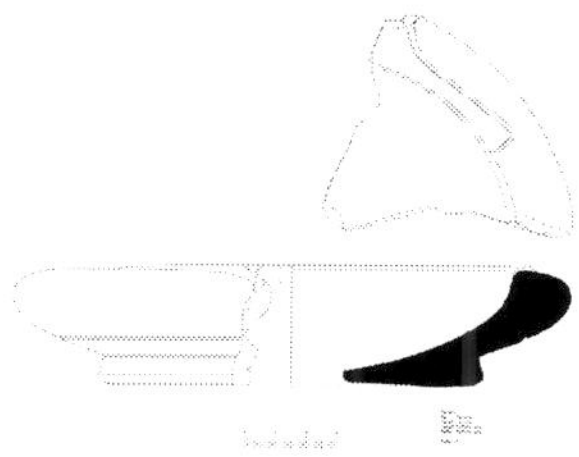

Figure 6. Mortar.

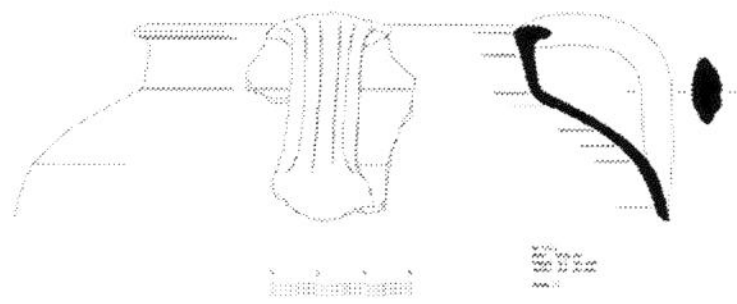

Figure 7. Pot/jar.

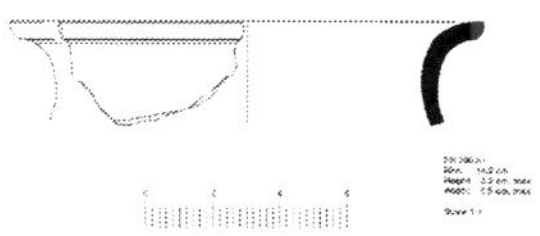

Figure 8. Jug.

Returning to the preliminary results, three general trends can be tentatively defined: (1) the kitchen wares of Greek tradition are predominant; (2) in most sites the casserole is the dominant shape used for cooking; (3) and in most large sites, the balance shifts so that pots are the dominant shape for cooking.

Examining these trends from the particular to the general, the first consideration is the variation in the shapes of the casseroles. The differences are limited mostly to details such as the curve of the wall or the size and inclination of the rim and lid-support shelf. The range of variation may be explained in a number of ways. It may, for example, represent a long period of production with development of shapes, or it may be the result of production by a large number of workshops or individual potters; it may even represent use of specific vessels for different cooking tasks. Because the number of sites representing short occupation is small (as of this writing), it is not possible to discard any of these possibilities as any or all of them may account for the variation. It is worthwhile, however, to recall that in the kiln deposit at Pantanello the pots, the dominant shape, demonstrate a similar pattern of minor but extensive variations in shape within a single period.

The discrepancy between use of open and closed cooking vessels on large and small sites is intriguing. So far there seems to be no correlation with location, nor with length of occupation, nor yet with the period represented. It seems most probable that the difference is related to the character of the site, and hopefully more robust suggestions can be made once all the evidence is assembled.

Finally, the question of acculturation returns. At the moment, few examples of the typically Roman kitchen shapes appear in the survey material, although the sample includes sites with occupation in the 3rd and even 2nd centuries BC, when Roman influence was pervasive. Again, interpretation has to be left open for the moment; before any conclusions are drawn, it is necessary to complete the study of the material in order to eliminate the possibility of biases in the sampling, and to correlate the kitchen wares with the remaining pottery. It is not premature, however, to suggest one or two possibilities. If there is a discrepancy between the finewares and the kitchen wares, it could be attributed to slow acculturation, and one must therefore look for patterns of influence. If not, one must consider different uses of the landscape between the periods.

In conclusion, the study of the kitchen wares has so far revealed a variety of interesting phenomena, which in turn will direct the method of future study. Some of these phenomena, such as the presence of large food processing installations and the absence of kitchen wares at certain sites, are certainly related to the character of a given site. Others suggest a number of hypotheses. Interpretation of these phenomena requires not only completion of the study of all kitchen wares from the survey, but also the integration of that study into the overall picture. While this is true for all categories of material culture—it is possible to reconstruct a picture of the life of the territory and the use of a landscape only through an integrated research methodology—it is particularly important in the study of kitchen wares, which, by virtue of the conservative

TERRACOTTA FIGURINES AND PLAQUES FROM THE PANTANELLO SANCTUARY AND METAPONTO FIELD SURVEY

By Rebecca Miller Ammerman
Colgate University

More than 20 years ago, as a graduate student, I made my first, brief visit to Metaponto. On that day I had the good fortune to encounter the Texas-style hospitality of Joe Carter and Ingrid Edlund-Berry, who graciously gave me a tour of the excavations in progress. Little did I imagine, as I listened to Prof. Carter explain (over the background din of the "well-point" pumps) the stratigraphy and history of the canalization of the spring at the Pantanello Sanctuary, that I would one day join the University of Texas team working on the publication of those excavations. In 2001, Prof. Carter invited me to join forces with Mary Malone in her study of the votive terracottas recovered during the course of excavations at the Pantanello Sanctuary, and also to assist with the publication of the terracotta figurines and plaques found in the field survey.

Mary and I were able to make two visits to Metaponto in 2002. The first visit took place in January; we arrived hours after Metaponto had seen its first snowfall in more than 50 years. I devoted the first few days to two basic procedures: (1) gaining an overview of the several hundred votive terracotta fragments found at the Pantanello Sanctuary and the work Mary had accomplished previously in her study of them and (2) beginning the study of the equal number of terracotta fragments recovered by the field survey. Each day Mary brought out case after case of fragments, patiently familiarizing me with the classifications that she had carefully worked out for the figurines and relief plaques unearthed at the Pantanello Sanctuary.

The practice of dedicating a baked clay statuette or plaque to a deity was a common ritual at ancient Greek sanctuaries. These terracotta objects were less expensive than statues in marble or precious metal, and thus they were affordable gifts for worshipers of modest means. Votive terracottas have been recovered in large quantities at most Greek sanctuaries in southern Italy. Each object represents the pious act of an individual who gave the offering in thanks for a service rendered by, or in the hopes of a future blessing from, the deity worshiped at that sanctuary. Moreover, the terracotta figurines and plaques contain a rich body of imagery (iconography) that was at some level considered appropriate for the cult at which the votives were dedicated. The interpretation of the iconography, especially in the absence of epigraphic or literary evidence specifying the patron deity of the sanctuary, is fundamental for understanding the nature of a given cult and the identity of the divinity that presided over it.

In the case of the Pantanello Sanctuary, the rite of dedicating a terracotta object comprised an important aspect of cult practice from the first half of the 6th century through the early 3rd century BC. Most of the figurines and plaques were formed in reusable molds, so that it is common to find several examples of a given image in the artifact assemblage. The earliest Archaic terracottas tend to represent a female figure wearing a polos, a tall conical headdress often interpreted as a marker of divine status, and a peplos, a simple body-length garment sometimes belted at the waist. (Fig. 1) Similar figurines, many of which were made from closely-related molds, have been recovered at other sanctuaries in the Metapontino, such as the one at San Biagio, as well as in the urban center itself. The excavations at Pantanello also yielded fragments of terracottas representing a female deity belonging to a popular group of Archaic figurines found at numerous sanctuaries connected to almost every Greek colony along the Ionian coast from Taranto to Croton, and at sanctuaries on the Tyrrhenian coast connected to Greek settlements such as Paestum and Palinuro. The sanctuaries containing these figurines have been associated with a wide range of deities. Thus, while the

Pantanello terracottas suggest worship of a female deity from the earliest period, the generic iconography of the figurines and the widespread occurrence of these types of terracottas prohibit a more specific identification.

The terracottas dedicated at Pantanello in the 5th and 4th centuries BC present similar challenges in the interpretation of their imagery. Several plaques portray a male figure reclining on a banquet couch, at the end of which sits a woman in a demure pose. (Fig. 2) The reclining figure may be a youth or an older, bearded man and usually wears a festive wreath embellished with palmettes, rosettes, and ribbons. The woman sometimes holds an infant, who stretches his arms out toward the reclining man. Occasionally the heads of two horses appear between the male and female figures to complete what seems to be a family portrait. The plaques are closely related to terracottas of reclining banqueters found in massive quantities at Taranto, ranging in date from the late 6th to the 3rd century BC. Similar plaques have been recovered at several locations within the urban center of Metaponto, at an industrial area near the its northern gate, and in the Metapontino at Crucinia, Sant'Angelo Vecchio, and several other sites as documented by the results of the field survey (see below). Various interpretations have been proposed for the identity of the male banqueter: a specific divinity (Dionysos or Hades) relaxing with his consort, or a deceased, heroized individual accompanied by family members. A chthonic, or funerary, connotation has often been proposed for the iconography of these plaques, but arguments have also been advanced suggesting that the plaques carried a political significance for the citizens of the colony. At Pantanello, the meaning that an image of a reclining banqueter may have held for a worshiper can be best assessed only once a detailed knowledge of the full assemblage of terracottas and other objects dedicated at the sanctuary has been attained, as well as a comparable knowledge about the presence of this specific imagery among the votives found at other Metapontine sanctuaries. This is clearly an important goal for future research.

This assessment is equally valid for another group of terracotta plaques found in abundance at the spring-sanctuary of Pantanello, dating from the late 4th to the early 3rd centuries BC. The details of the plaques vary, but all portray a satyr-like male figure frolicking with a young female in a rocky landscape festooned with clusters of grapes; the figures are usually identified as Pan with a nymph, or a Silenos with a maenad. (Plate 9b, p.46) In either case, the imagery evokes the sphere of Dionysos. Closely related plaques have been found at Temple E and elsewhere in the urban center of Metaponto, as well as at several sites in the chora. It is noteworthy that at several of these sites the plaques occur in association with those portraying a reclining banqueter.

In January I also examined some 400 fragments of terracottas collected during the course of the field survey. With the assistance of archaeological photographer Giambattista Sassi, I brought dozens of boxes overflowing with survey artifacts from the storage shelves down to the work tables in the "Banca" to search for terracotta fragments. Giambattista photographed any fragment that I recognized as belonging to a specific type, while I made quick notes on the terracottas from each of the sites.

In July I returned to Metaponto for 16 days with Mary Malone and Bryan Kraemer, an American student currently pursuing an M.Sc. in archaeological computing at the University of Southampton who had assisted me with the study of the terracottas at Paestum

Figure 1. Archaic terracotta statuette of a standing female figure from the Pantanello Sanctuary. (Photo: G. Sassi).

in 2001 through the American Academy in Rome's Summer Program in Archaeology. In the spring I had begun construction of a database for the Pantanello terracottas as well as another database for the survey terracottas, inserting the digital photographs that had been taken in January by Giambattista. I had also refined the preliminary typologies developed by Mary Malone for many of the Archaic figurines and reclining banqueters from Pantanello, allowing us to begin sorting out typologies for the more complex set of fragmentary plaques of satyrs and maenads in the summer. More than 400 terracottas from the sanctuary have thus far been selected for inclusion in the database, which will serve as the basis of the published catalogue.

Significant progress was also made on the compilation of the survey terracotta database. Ninety-seven sites have yielded 388 fragments of terracotta figurines or relief plaques. Almost half of these fragments can be identified by subject and date. Plaques of reclining banqueters or of satyrs and maenads are the most common, represented by more than 30 examples each. Of the eight sites where ten or more terracotta fragments were collected, the iconography of satyrs or maenads is present at six; and at three of these six sites, the satyr/maenad plaques are associated with plaques of reclining banqueters. In other words, a pattern is emerging, also found at the Pantanello sanctuary, that may be typical of religious activity in the Metapontino in the 4th and early 3rd centuries BC. The two remaining sites at which numerous terracotta fragments were collected arouse interest for the prevalence of Archaic figurines in the assemblages.

While much research remains to be done, the preliminary documentation of the terracottas from the field survey is already beginning to sketch a fuller picture of the religious articulation of Metaponto's rural landscape.

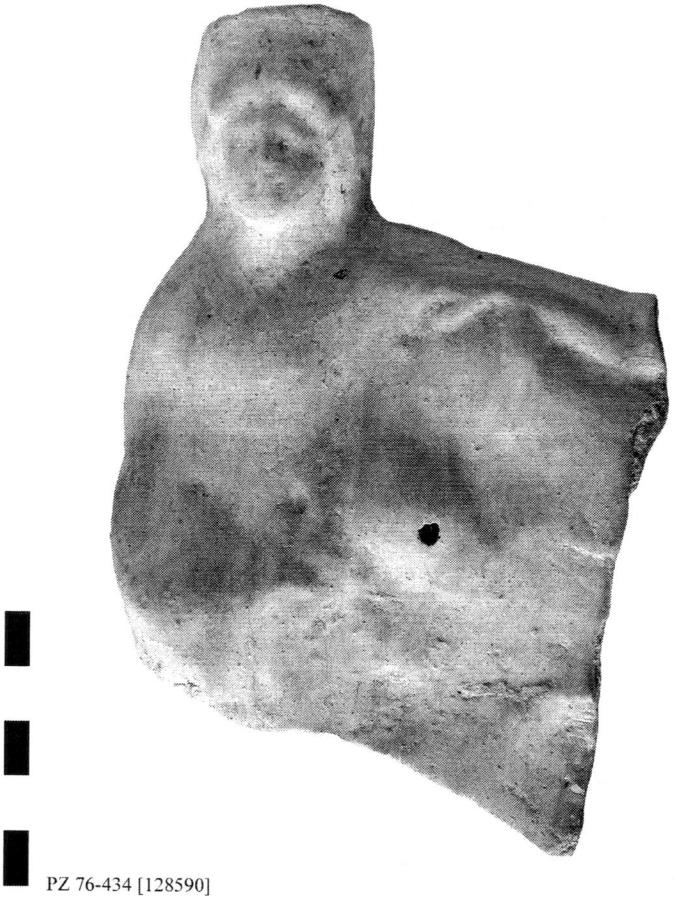

PHOTO: GIAMBATTISTA SASSI

Figure 2. Terracotta fragment from the Pantanello Sanctuary depicting a woman seated beside a banqueter at the end of a couch. (Photo: G. Sassi).

The Greek and Roman Amphorae from the Metaponto Field Survey

By Marsha Robbins

University of Southampton, England

In studying the large undecorated storage vessels, the amphorae, I am attempting to reveal the basic routes of exchange within the chora of Metaponto during the periods of Greek and Roman occupation, which will also provide evidence for probable systems of exchange in the neighboring territories and regions. One focus of this study is the impact of Roman control on the Metapontino and the other Greek territories of Magna Graecia. Preliminary results strongly suggest that the Roman occupation had a negative effect on the regional economy. The date-ranges of the sites containing amphorae are indicative (Fig. 1): while sites containing amphorae are numerous between the 7th and 3rd centuries BC, when Greek influence on the territory is strong, after the 3rd century BC the number of sites containing amphorae decreases notably. It was at this time, circa 280 BC, that the Romans gained complete control of southern Italy.

Fig. 2 lists the specific types of amphorae found at each site, with their dates of manufacture. Here we see quite clearly high percentages of imported amphorae, dating between the 8th and 3rd centuries BC, probably indicating that the inhabitants of the chora had sufficient funds to support the purchase of imported wines and oils from elsewhere in the Greek world. (Fig. 3) Locally-manufactured amphorae of this period, by contrast, are very few. It was not until the mid-2nd century BC that amphorae made of local fabrics were introduced into the territory, with the subsequent disappearance of virtually all imports until the 4th century AD. (Fig. 4) These findings strongly suggest that the economy of the territory suffered a severe contraction early in the 2nd century BC, which forced the populace to rely heavily on their own produce, so that the importation of other (and perhaps better) products was no longer an option.

The distinction between local and non-local fabrics lies in the clay matrix and the inclusions (e.g., sand or mineral particles) it may contain—its "ingredients," so to speak. Most matrices have distinct combinations of minerals and inclusions that can be geologically traced to specific regions. Only thin-section analysis can accurately identify where the different clay fabrics, and thus the different amphorae, originated. By slicing very thin

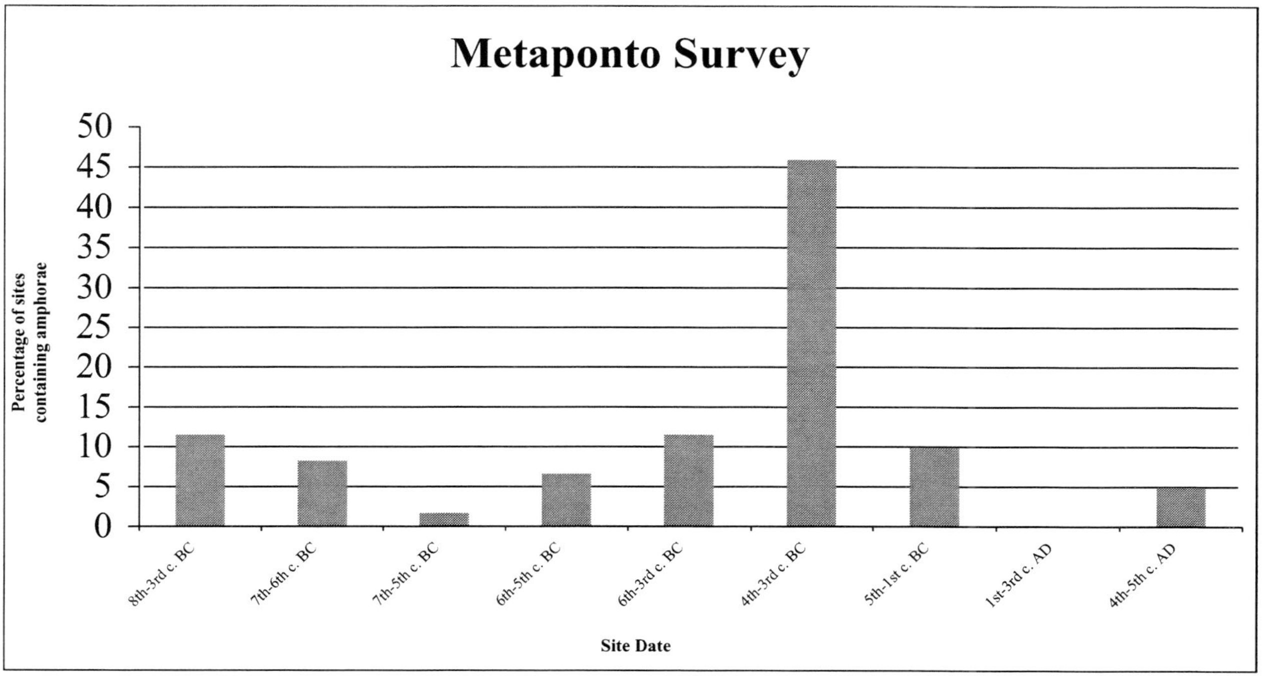

Figure 1. Percentages of sites in the Metapontino containing at least five amphora fragments (total of 61 sites), with corresponding date ranges. The date ranges given on the x-axis are based on the amphorae found at each survey site. The y-axis shows the percentage of sites which fit into these date ranges.

(30–40 microns) sections of a ceramic fragment one can see—with the aid of a microscope—each particle of quartz, mica, feldspar, etc. in the fabric. This part of the analysis is still a work-in-progress, so that much of the fabric identification to-date is based on coarse visual analysis with a hand-held magnifying glass (10x power), which is nevertheless a satisfactory tool for answering the broad question of local versus non-local provenience.

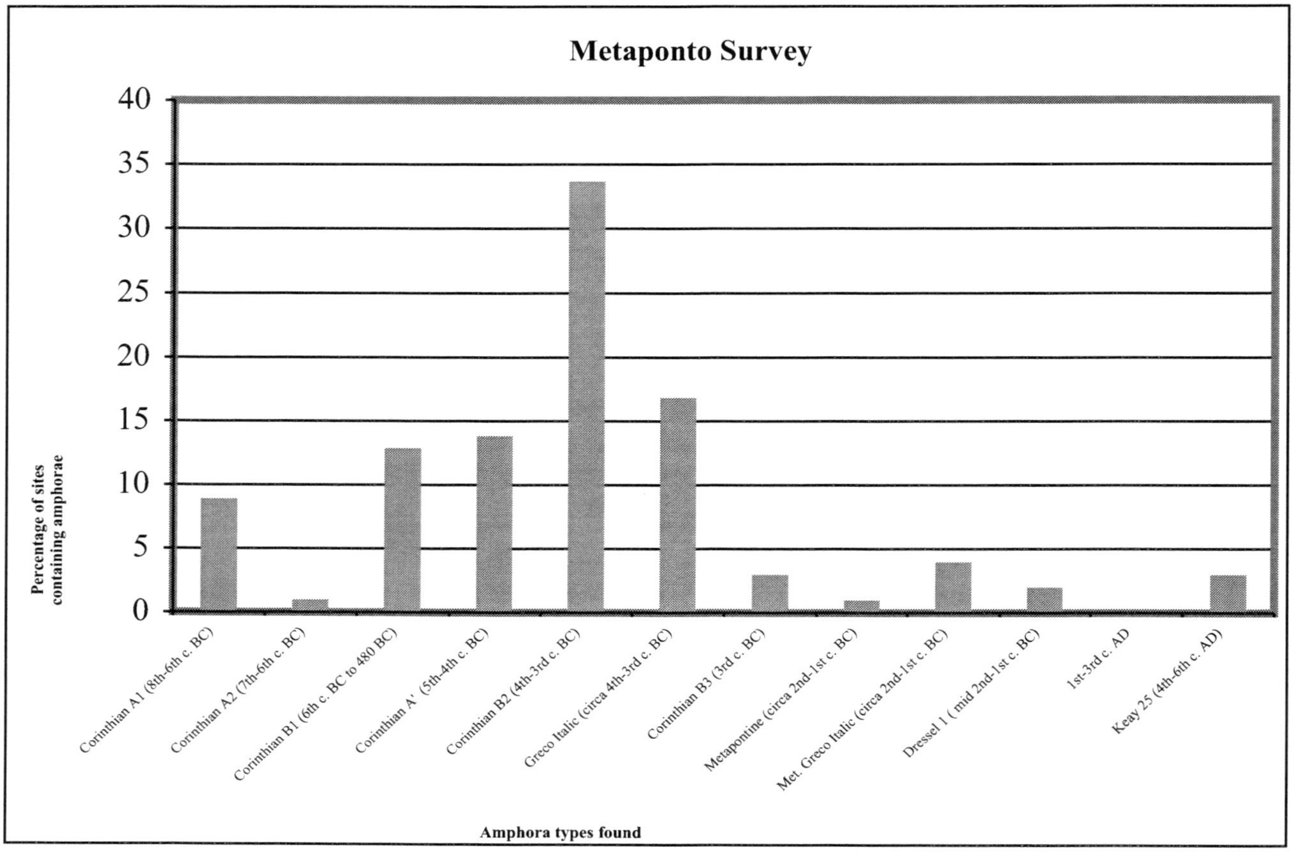

Figure 2. The different types of amphorae found in survey sites in the Metapontino containing five or more amphora fragments (total of 61 sites). On the x-axis: amphora types with corresponding dates of manufacture. The y-axis: percentages of sites in which the types appear. Only one amphora of each type was counted at each site.

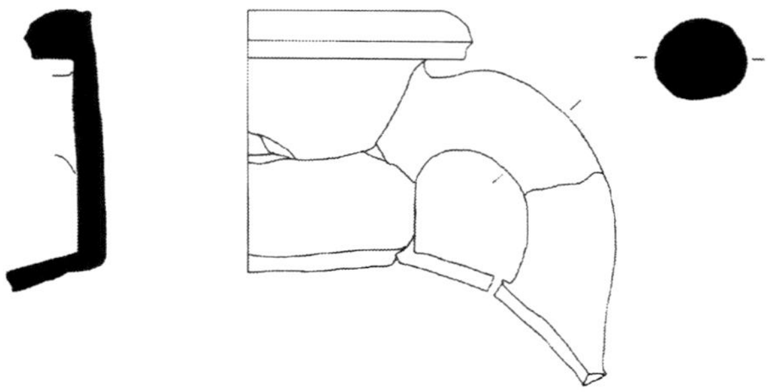

Figure 3. Corinthian type A2 amphora, 7th–6th c. BC

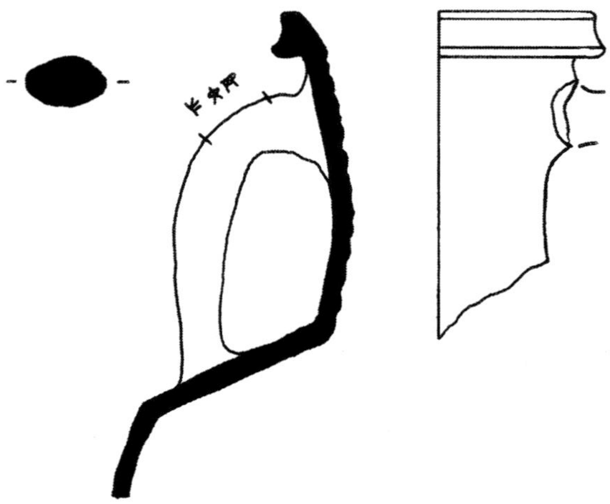

Figure 4. Metapontine (Greco-Italic) amphora, 2nd–1st c. BC.

Black Gloss Pottery from the Pantanello Sanctuary: A Progress Report

By Mariah Wade
University of Texas at Austin/ICA

Introduction
My companions in the Pantanello Sanctuary ceramics project frequently complain about the length of time required to produce and then evaluate the results of our labors. The work is intensive, the progress slow, and it is difficult to explain to anyone but a ceramicist why such small fragments of pottery demand so much study and so many resources. Although we are still far from the end, there is light at the end of the tunnel.

In the last year, the members of the ceramic study group have digitized and cleaned approximately 450 drawings of vessels (black gloss and coarsewares, as well as loomweights) for inclusion in the forthcoming publication of the Pantanello Sanctuary excavation (Figs. 1–2). In the process of creating them, UT students Rosalyn Mandola, Matt Hibbs, and Thomas Howe overcame many practical difficulties to produce clean, informative, and visually pleasing reproductions of real objects. There are still many drawings to be digitized and cleaned, particularly those made early in the life of the project (1970s-1980s), but good progress has been made.

The Publication
As project leader, I have completed the central typological study and identified the crucial comparative pieces for various groups of the black gloss vessels. UT journalism student Brian Wellborn is currently creating mock computer layouts of the text, drawings, and photographs for individual groups of vessels in order to determine the optimum page size, image scale, and distribution of text and images (Fig. 3). At the same time he is evaluating text readability (font size and style), the problems of depicting different sizes of vessels on the same page, and other general issues in the presentation of the material.

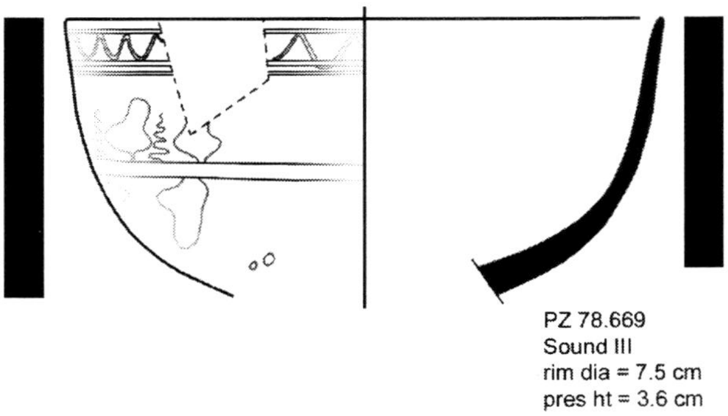

Fig. 1: Drawing of black gloss bowl with ovule and ivy wreath decoration.

Ceramicists' work is highly specific and detailed, particularly when dealing with large and fragmentary assemblages. They have to research similar fragmentary pieces from other archaeological sites in order to identify and date their own materials, and the publication of their results requires a clear descriptive text, as well as drawings and photographs that unequivocally communicate the basic diagnostic features of the fragments. Preparation of these elements is tedious and often frustrating work, as demonstrated by many publications in which the drawings are too small, or they are "artistic" without being informative, the scale is absent or indiscriminately applied, and the photographs are of poor quality or entirely absent. Painfully aware of these common shortcomings, we have created the mock layouts to help us identify and correct many such pervasive problems and other serious issues before reaching the final phase of the project.

The importance of producing a publication that is user-friendly for ceramicists who have to perform similar research work, and at the same time maintains high scholarly and visual standards, is clear. In the process of devising effective methods to present and use the ceramic material, we have also become aware of the need for a publication that can be used as a teaching tool. Several features, still being elaborated and evaluated, will be included in the finished volume to make it both a scholarly reference for professional archaeologists investigating the Greek ceramics of southern Italy and a tool for teaching the subject at the undergraduate and graduate levels. These features will extend the publication's utility well beyond the limitations of a traditional archaeological site report and ceramics

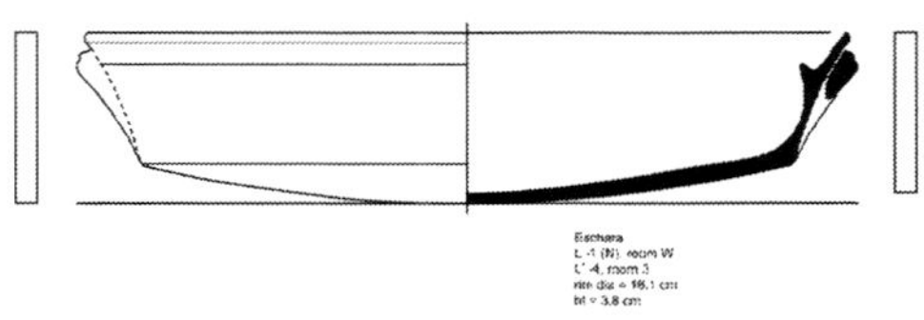

Fig. 2: Drawing of cooking casserole (lopas).

research tool (as it was originally conceived), creating what we hope will become a model for future studies of ancient ceramics, thus enhancing its humanistic value enormously.

Except for the actual printing and binding, all work on the publication will be done "in-house" by ICA employees; the final version will be presented to the publisher as a camera-ready product on disk, as was its oft-cited predecessor in the *Chora of Metaponto* series, *The Necropoleis* (University of Texas Press, 1998). This is the most satisfactory method for producing an archaeological report, since each stage is controlled by archaeologists and persons intimately familiar with the subject matter. It also reduces the cost of publication and provides needed employment and valuable professional experience for UT graduate and undergraduate students (although the students' working time must always be subordinated to academic responsibilities). This solution has proven to be reciprocally rewarding: the students' enthusiasm, engagement, and unbiased thinking have provided fresh and innovative solutions to old problems. As resources for research and publication continue to contract, especially in the humanities, this economical method of increasing the body of new, fundamental knowledge should offer hope to aspiring scholars. The role of ICA and the Packard Humanities Institute is essential for the dissemination of substantive basic research that rises above prevailing academic trends.

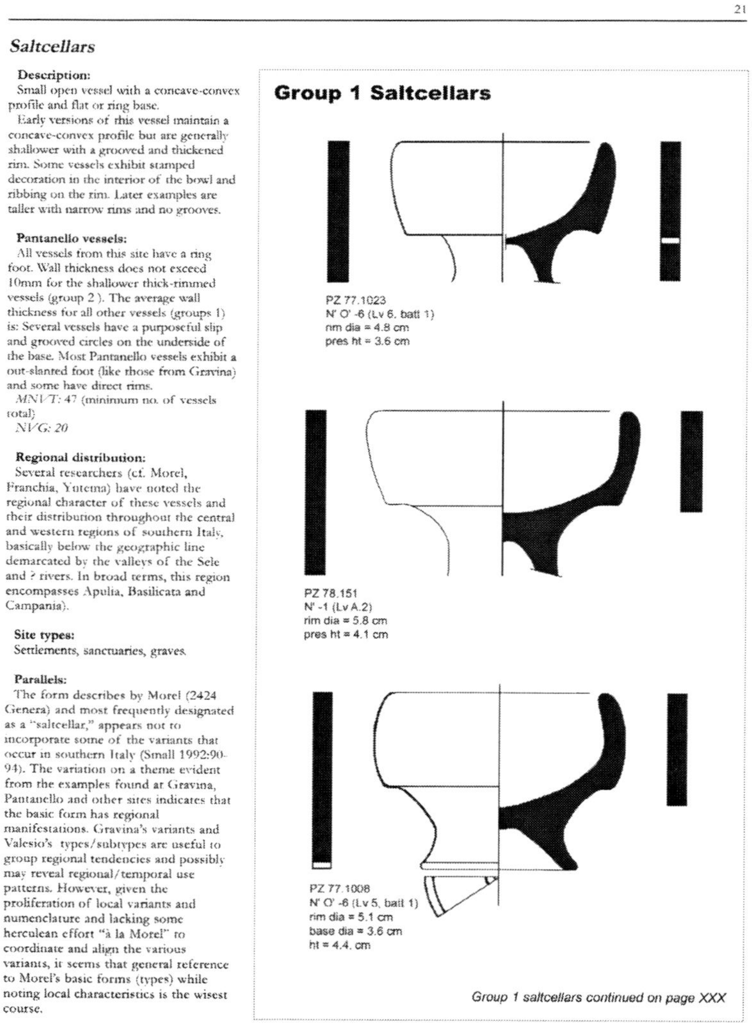
21

Saltcellars

Description:
Small open vessel with a concave-convex profile and flat or ring base.
Early versions of this vessel maintain a concave-convex profile but are generally shallower with a grooved and thickened rim. Some vessels exhibit stamped decoration in the interior of the bowl and ribbing on the rim. Later examples are taller with narrow rims and no grooves.

Pantanello vessels:
All vessels from this site have a ring foot. Wall thickness does not exceed 10mm for the shallower thick-rimmed vessels (group 2). The average wall thickness for all other vessels (groups 1) is: Several vessels have a purposeful slip and grooved circles on the underside of the base. Most Pantanello vessels exhibit a out-slanted foot (like those from Gravina) and some have direct rims.
MNVT: 47 (minimum no. of vessels total)
NVG: 20

Regional distribution:
Several researchers (cf. Morel, Franchia, Yntema) have noted the regional character of these vessels and their distribution throughout the central and western regions of southern Italy, basically below the geographic line demarcated by the valleys of the Sele and ? rivers. In broad terms, this region encompasses Apulia, Basilicata and Campania).

Site types:
Settlements, sanctuaries, graves.

Parallels:
The form describes by Morel (2424 Genera) and most frequently designated as a "saltcellar," appears not to incorporate some of the variants that occur in southern Italy (Small 1992:90-94). The variation on a theme evident from the examples found at Gravina, Pantanello and other sites indicates that the basic form has regional manifestations. Gravina's variants and Valesio's types/subtypes are useful to group regional tendencies and possibly may reveal regional/temporal use patterns. However, given the proliferation of local variants and numenclature and lacking some herculean effort "à la Morel" to coordinate and align the various variants, it seems that general reference to Morel's basic forms (types) while noting local characteristics is the wisest course.

Fig. 3: Preliminary publication layout for black gloss saltcellar groups.

ACTIVITY AT THE CENTER FOR AGROARCHAEOLOGY AT PANTANELLO

Rosetta Torraco
[Translated by Alberto Prieto]

In March of 2000 the Center for Agroarchaeology (Centro di Agroarcheologia) at Pantanello was born, following the signing of an agreement between the Department of Agriculture and Forestry of Basilicata and ICA. This formalization came about after nearly three decades of informal collaboration between the two institutions, thanks to which ICA has received hospitality and logistical support from the regional government of Basilicata in the structures of the Azienda Sperimentale at Pantanello.

According to the agreement, ICA obtained the use of two structures, which it agreed to renovate within five years. In 2000-2001 the renovation activity was directed towards the empty two-story house that had in the past lodged the family of the Azienda's foreman. (Fig. 1) The bedrooms, bathroom, and kitchen, located on the second floor, were refurbished to house guests of the Center; the three storage rooms on the first floor were renovated to serve as offices and restructured with the addition of a full bathroom for the Center staff, ICA guests, and field crews. The general renovation work consisted of exterior replastering, interior painting, repair of damaged doors and windows (along with shutters and screens), replacement of damaged floor surfaces, update of electrical systems to meet current specifications, and installation of telecommunications lines and climate control.

During 2001 and the first half of 2002 the first-floor offices were equipped with two computers, an inkjet printer, an image scanner, a local area network file-sharing connection, Internet access, a fax machine, a telephone answering machine, a digital camera, chairs, tables, and storage shelves. The second-floor residence was equipped with pots and pans, refrigerator, clothes washing machine, reading lamps, beds, bookshelves, chairs, and tables.

In 2002 the renovation activity was directed towards the creation of a library/reading room and two stud-

Figure 1: The Pantanello house after renovation in late 2000. (Photo: J. Carter)

ies in a one-story structure of approximately 120m^2 (1300ft^2) constructed near the beginning of the last century. This structure was in significantly worse shape; the renovation required the complete replacement of the roof and exterior/interior plaster (which was thoroughly rotted by water damage), repair of damaged floor surfaces, reactivation of a long-unused bathroom, and repair and repainting of the existing wooden fixtures.

At the end of 2002 the preparation of the library/reading room began. (Fig. 2) This space will also serve as the central study-room and conference room, accommodating small seminars and professional meetings. The current phase of work involves the creation of the library's wall-mounted wooden shelves. In 2003 work on this structure will include an update of the electrical system as well as installation of telephone lines, a local area network, and climate control. Also in progress is the creation of a database cataloguing all of the written materials at the Center that pertain to past and present ICA studies and research. A complete database of all books and scholarly articles in the Center's library has already been created. In 2003 and following years the library will be significantly enriched through acquisition of scholarly publications, making the Center an important cultural feature of southern Italy.

In 2002 the Center hosted many scholars and specialists: in January Mary Malone and Prof. Rebecca Ammerman; in May Profs. Renata and Maciej Henneberg; in June Marsha Robbins; in July Mary Malone and Prof. Ammerman again, along with Prof. Albert Ammerman; in September, Smadar Gabrieli, Richard Fletcher, Cesare D'Annibale, Andrei Opait, Marsha Robbins, and Michele Roccia; throughout the year, archaeologists and artists Cesare Raho and Eloisa Vittoria, as well as graduate students from the various universities at which the professors work. Finally, the Center and its excavated sites at Pantanello have been the object of numerous guided tours by scholastic groups, scholars, and tourists of many nationalities.

Figure 2: The interior of the library/reading room at Pantanello in late 2002, during renovation and installation of the shelves. (Photo: R. Torraco)

Chersonesos
2002 Campaign

Professor Joseph Coleman Carter, Director, ICA
Taissa Bushnell, MA, Director, *Pidtrymka Chersonesu*

ICA Administration in Austin
Carol K. Cook, MA
Pat Irwin

Kyiv Coordinator
Professor Svetlana Telenkova

National Preserve of Tauric Chersonesos (NPTC)

Dr. Leonid Marchenko, General Director
Dr. Galina Nikolaenko, Vice Director for Research
Dr. Larissa Sedikova, Vice Director for Conservation
Ludmila Grinenko, Head Librarian
Stanislav Ryzhov, Head of Architecture & Archaeology

Excavation, Ancient City

Professor Paul Arthur, Co-Field Director, University of Lecce, Italy
Dr. Larissa Sedikova, Co-Field Director, NPTC

Excavation team
The Univ. of Kyiv-Mohyla Academy, Kyiv
Professor Vitaly Zubar, supervisor
Pavlo Nechytaylo
Oksana Petik
Tetiana Starostenko
Yaroslav Sukhomlinov
Olha Timchenko
Petro Yevchyn

Taurian National University, Simferopol
Nikita Khrapunov
Danila Kostromichev
Alexander Kuzmin
Georgiy Tsarevski

Shevchenko National University, Kyiv
Kateryna Motsya
Larissa Novytska
Mykhaylo Pryshchepa, Culture University, Kyiv

Mikhail Fomin, Kharkiv School of Arts
Marco Leo Imperiale, Univ. of Lecce
Mykhaylo Pryshchepa, Nat'l Univ. of Culture, Kyiv
Adela Sobotkova, Masaryk Univ. of Brno, Czech Rep.

Site documentation and supervision
Brunella Bruno, Univ. of Lecce
Marco Leo Imperiale, Univ. of Lecce
Danila Kostromichev, TNU., Simferopol
Marina Montano, Univ. of Lecce

Site GIS
Jessica Trelogan, MA, ICA
Alessandro Rizzo, Univ. of Lecce

Photography
Adela Sobotkova
Chris Williams, ICA

Database
Giuseppe Gravili, Univ. of Lecce

AutoCAD and reconstructions
Massimo Limoncelli, Univ. of Lecce

Finds laboratory
Paola Tagliente, Univ. of Lecce, supervisor
Olga Andreeva, NPTC
Tatiana Borodinskaya
Giuditta Cavallo, Univ. of Lecce
Irina Don, NPTC
Irina Gusakova, NPTC

Coin identification
Ludmila Golovchenko, Head of the Collections, NPTC
Elena Kochetkova, Collections, NPTC

Administration and logistics in Lecce
Salvatore De Vanna
Brunello Bruno

Excavation, Chora

Dr. Galina Nikolaenko, Head of Heraklean Expedition, NPTC
Adam Rabinowitz, Co-Field Director, University of Michigan
Tatiana Yashaeva, Co-Field Director, NPTC

Excavation team

The University of Kyiv Mohyla Academy, Kyiv:
Oleh Dubinets
Andriy Ivchenko
Evelina Kravchenko
Dmytro Peftits
Dmytro Shevchenko
Olha Vergun

Gordon Lobay, Ph.D candidate, Univ of Cambridge
Lauren Aicklen, UT
Ethan Spanier, University of San Francisco
Laura Surtees, Canadian Institute, Athens

Ceramics

Dr. Andre Opait, Toronto
Anca Dan, Sorbonne, Paris
Antonina Shevchenko, NPTC
Adela Sobotkova
Denis Zhuravlev, State History Museum, Moscow

Remote Sensing and GIS

Jessica Trelogan, MA, ICA
Jamon van den Hoek, University of Wisconsin

NPTC Conservation Laboratory

Inga Shvedova, Laboratory Head
Olga Demyanova
Julia Ryzhova
Natalia Kuleva

Polychrome Stelai Project

Prof. Joseph Carter, ICA
Prof. Richard Posamentir, Univ. of Bochum, Germany
John Twilley, Conservation Scientist, Hawthorne, NY
Chris Williams, ICA
Evgeniy Zherebtsov, Sevastopol
Prof. Paula Perlman, Department of Classics, UT
Prof. Alla Buyskykh, Institute of Archaeology, Ukrainian National Academy of Sciences
Jodi Lane, Architect, Austin

Museum Computerization and Archival Preservation (Megarika Project)

Glenn Mack, MA, ICA
Ludmila Grinenko, Head Librarian, NPTC
Oleg Vertolitsky, System Administrator, Pidtrymka Chersonesu
Nonna Krasovskaya, Head Archivist, NPTC
Tatiana Dianova, Archivist, NPTC
Alexander Semenov, Team Leader, Pidtrymka Chersonesu
Andrei Kondyuk, Lead Scanner, Pidtrymka Chersonesu
Alexei Romanov, System Administrator, Pidtrymka Chersonesu

Palaeobotany, Paleoecology, and Palynology

Prof. Carlos Cordova, Oklahoma State University Stillwater, Oklahoma
Dr. Galina Pashkevich, Institute of Archaeology, Ukrainian National Academy of Sciences
Anna Zhukova, Taurian National Univ., Simferopol

Anthropology

Vladimir Smirnov, Crimean Bureau of Forensic Medicine, Simferopol
Denis Ponomarev, Crimean Bureau of Forensic Medicine, Simferopol

Archaeozoology

Dr. Alexei Kasparov, St. Petersburg

Geophysical Prospection

Mikhail Nikolaenko, Head of Technical Department, NPTC
Sergei Shakuro, Nizhny Novgorod

Planning and Architecture

Tatiana Bazhanova, Architect, State Research Institute of Architecture and Urban Planning, Kyiv
Nikolai Andrushchenko, Architect, State Research Inst. of Architecture and Urban Planning, Kyiv
Alma Maldonado Holiday, Landscape architect, Austin
Carl Holiday, Architect, Austin
Jodi Lane, Architect, Austin
Olga Ushakova, St. Petersburg, Russia

Conservation Consultants

Chris Cleere, Cleere Conservation, Ltd., UK
Semih Ercan, Cleere Conservation, Ltd., UK

CONSERVATION OF SITE 132

Vera Nikolaenko, Leader, NPTC
Lilia Bondareva, Nikitski Botanical Garden, Yalta
Olga Grigorieva, Sevastopol
Alexander Gusakov, Sevastopol
Sergey Kobelev, Sevastopol
Anna Kovaleva, Sevastopol
Sergey Kutykin, Sevastopol
Olga Mochalkina, Sevastopol
Svetlana Prut, Sevastopol
Tatiana Sivochub, Sevastopol
Roman Smirnov, Sevastopol

PUBLICATION OF *CRIMEAN CHERSONESOS*

Glenn Mack, MA, ICA
Ludmila Grinenko, Head Librarian, NPTC
Nikita Khrapunov, Taurian National University, Simferopol

WEBSITE (WWW.CHERSONESOS.ORG)

Viktor Mussuri, *Pidtrymka Chersonesu*

TRANSLATION AND INTERPRETING

Irina Fadeeva, Sevastopol
Marina Krotova, Sevastopol
Natalia Streltsova, Sevastopol
Professor Svetlana Telenkova, Kyiv

STORAGE AND LABORATORY CONSTRUCTION

Yuri Aliabev, Sevastopol director, Krymrestauratsiya
Ludmila Dukareva, independent architect, Sevastopol
Alexander Levytsky, independent engineer, Sevastopol

LOCAL LOGISTICAL SUPPORT

Viktor Streltsov, Vice Director for Operations, NPTC
Tatiana Gagarina, NPTC
Ludmila Khudyakova, Sevastopol
Oleg Oktyabrev, NPTC
Oksana Sukach, Sevastopol
Svetlana Sukach, Sevastopol
Raissa Tychinskaya, NPTC

TRANSPORTATION AND SUPPLIES

Vasily Kirichenko, Kyiv
Sergei Koseikin, Simferopol
Nikolai Prut, NPTC
Pavel Statsenko, NPTC
Orest Yavtushenko, Kyiv

ACCOUNTING, LEGAL, AND BANKING

Elena Baryshnikova, MBA, Lex Service, Sevastopol
Ekaterina Kruchinina, Lex Service, Sevastopol
Elena Filipova, Lex Service, Sevastopol
Grigoriy Lutsai, Director, Sevastopol Branch, State Import-Export Bank of Ukraine

SENIOR ADVISORS

Prof. Aleksandr Aibabin, Taurian National University, Simferopol
Prof. Neal Ascherson, Editor, Public Archaeology, London
Yuri Bohutsky, Minister of Culture and Arts of Ukraine
Prof. Henry Cleere, World Heritage Coordinator, ICOMOS, United Kingdom
Evnika Linyova, Department for the Preservation of Cultural Heritage, Ministry of Culture and Arts of Ukraine
Ambassador William Green Miller, Former US Ambassador to Ukraine, 1993-1998
Leonid Novokhatko, Deputy Secretary, Ministry of Culture and Arts of Ukraine
Ambassador Carlos Pascual, U.S. Ambassador to Ukraine
Academician Petro Tolochko, Director, Institute of Archaeology, Vice President of the Ukrainian National Academy of Sciences
Prof. Leonid Zaliznyak, the University of Kyiv Mohyla Academy
Evgeniy Zherebtsov, Sevastopol
Prof. Vitaly Zubar, The University of Kyiv-Mohyla Academy

METAPONTO
2002 Campaign

CERAMICS STUDY

Professor Mariah Wade, Director of Ceramic Studies, UT Anthropology and ICA
Prof. Rebecca Miller Ammerman, Colgate University, Terracotta figurines
Vincenzo Cracolici, Bari, Black gloss pottery
Richard Fletcher, Ph.D. candidate, University of Sydney, Databases
Smadar Gabrieli, Ph.D. candidate, University of Sydney, Cookingware
Erminia Lapadula, University of Lecce, Late Roman and Medieval pottery
Mary L. Malone, MA, Austin, Terracotta figurines
Dr. Cesare Raho, Metaponto, Illustrator
Marsha Robbins, Ph.D. candidate, University of Southampton, Amphorae
Dr. Anna Lucia Tempesta, Metaponto, Architectural terrracottas
Dr. Eloisa Vittoria, Metaponto, Black gloss pottery
Don Wade, MSc, Austin, Geologist
Dr. Bronwen Wickkiser, Ph.D., UT Classics, Loomweights

UT Austin Students

James Collins, UT, MA candidate, Stanford University, Illustrator
Matthew Hibbs, UT, Computer Sciences, Databases
Thomas Howe, MA candidate, UT School of Government, Databases
Rosalyn Mandola, BA, UT French and Italian, Databases
Melinda Spearman, MA, UT Classics, Unguentaria
Brian Wellborn, UT, Journalism, Layout & Design
Ting Ting Wu, Ph.D. candidate, UT Mathematics, Statistics

SURVEY PUBLICATION

Prof. Joseph Coleman Carter, ICA, co-editor
Alberto Prieto, Ph.D. candidate, UT Classics, ICA, co-editor
Dr. James Abbott, Texas Department of Transportation, Austin
Cesare D'Annibale, Consultant, Parks Canada
Prof. Robert L. Folk, UT Geology
Dr. Stephen Thompson, ICA

PHYSICAL ANTHROPOLOGY

Prof. Maciej Henneberg, University of Adelaide
Dr. Renata Henneberg, University of Adelaide

2002 Publications by ICA Staff Members and Collaborators

Ammerman, Rebecca Miller. *The Sanctuary of Santa Venera II. The Votive Terracottas.* Univ. of Michigan Press, Ann Arbor. 2002.

Carter, J.C. "Color at Chersonesos (on the Black Sea): Funerary Monuments from the Early Hellenistic Necropolis." In Tiverios, M.A. and D.S. Tsiafakis *Color in Ancient Greece. The Role of Color in Ancient Greek Art and Architecture.* 161–170. Thessaloniki, 2002 .

Twilley, J. "Pigment Analyses from the Grave Stele and Architectural Fragments from Chersonesos." *Color in Ancient Greece. The Role of Color in Ancient Greek Art and Architecture.* 171–178. Thessaloniki, 2002 .

The Study of Ancient Territories: Chersonesos and Metaponto. 2001 Annual Report. Institute of Classical Archaeology, Austin, 2002. (ISBN 0-9708879-1-4).

2002 Publications about the Metaponto and Chersonesos Projects

Ascherson, N. "Chersonesos: Public Archaeology on the Black Sea Coast." *Archaeology International,* 54–56. Institute of Archaeology. London, 2002.

Ridgeway, D. "Archaeology in Sardinia and Southern Italy, 1995–2001." *Archaeological Reports, 2001–2002,* 130–134.

Romey K. M. "Legacies of a Slavic Pompeii." *Archaeology.* 55, 6:18–25.

Snodgrass, A. A. "A Paradigm Shift in Classical Archaeology?" *Cambridge Archaeologial Journal.* 12: 2, 179-194. Cambridge, England, 2002.

World Monuments Watch 2002. 100 Most Endangered Sites. 83. World Monuments Fund, New York,

Invited Papers at Meetings by ICA Staff Members and Collaborators

Carter, J. C. Chersonesos. "A Joint Ukrainian–U.S. Archaeological Project." Inaugural lecture as a member of the Brotherhood of Kyiv'Mohyla Academy University, Kyiv, June 25, 2002.

———. "L'ambiente ed il Paesaggio a Metaponto." 42° Convegno internazionale di studi sulla Magna Grecia, Matera, Italy, Oct. 7, 2002.

———. "The Contributions of Science (Physical, Biological) to the discovery of Ancient Rural Civilization." Annual Fall Meeting, American Philosophical Society, Philadelphia, Nov. 9, 2002.

Carter, J. C., L. Burn, and M. and R. Henneberg. 2002. "Burial in the Chora of Metaponto." 103rd Annual Meeting of the Archaeological Institute of America, Philadelphia, Jan. 4. 2002.

Cordova, Carlos and Paul Lehman. "Paleoecological Research in Southwestern Crimea." Eurasian Archaeology Conference, The University of Chicago, May 3–4, 2002

Opait, A. "The Eastern Mediterranean Amphorae in the Province of Scythia." International Colloqium at the Danish School in Athens, September 26–29, 2002, Athens.

———. "'Table' Amphora and 'Table' Pitcher." Rei cretariae Romanae Fantores 23rd International Congress, Rome, October, 2002.

Sedikova, L., P. Arthur, and S. Ryzhov. "Excavation at Chersonesos, 2001." AIA, Philadelphia, Jan. 6, 2002.

Trelogan, J., M. Crawford, and J. C. Carter. "Monitoring the Ancient Countryside: Remote Sensing and GIS at the Chora of Chersonesos (Crimea, Ukraine)." Abstract publication, *Proceeding of the Space Applications for Heritage Conservation Conference.* Strasbourg, France. Nov. 5, 2002.

Trelogan, J. and G. Nikolaenko. "The Chora of Chersonesos in Crimea in the Early Hellenistic Period." AIA, Philadelphia, Jan. 6, 2002.

Thompson, S. M. "The Metaponto Archaeological Survey: The 2001 Field Season: Resurvey and Longitudinal Study of the Surface Archaeological Record." Poster session, AIA, Philadelphia, January 3–6, 2002.

———. "Side-by-Side and Back-to-Front: Intra-Regional Latitudinal and Longitudinal Comparability in Survey Data. Three Case Studies from Metaponto, Southern Italy." Side-by-Side Survey: A Workshop in Comparative Regional Studies in the Mediterranean World, Ann Arbor, Michigan, April 5-7, 2002.

———. "Colonization in Context: The Cultural Geography of Pre-Greek Eastern Sicily." 67th Annual Meeting of the Society for American Archaeology, Denver, CO, March 20–24, 2002.

COPIES TO DONORS AND:

The Honorable Yuri Bohutsky, Minister of Culture and Arts of Ukraine

The Honorable Kostyantyn Gryshchenko, Ukrainian Ambassador to the U.S.

The Honorable William Green Miller, former U.S. Ambassador to Ukraine

The Honorable Carlos Pascual, U.S. Ambassador to Ukraine, Kyiv

The Honorable James D. Wolfensohn, President, The World Bank, Washington, D.C

Ms. Lisa Ackerman, The Samuel H. Kress Foundation, New York

Professor Aleksandr Aibabin, Taurian National University, Simferopol

Ms. Bonny Burnham, President, World Monuments Fund, New York

Dr. Antonio De Siena, Direttore, Museo Archeologico Nazionale, Metaponto

Ms. Janet Demiray, Public Affairs Officer, US Embassy, Kyiv

Mr. Paul Hacker, Cultural Affairs Officer, US Embassy, Kyiv

Mr. Destry Jarvis, Private Consultant on Cultural Resource Management

Dean Michael Katz, Middlebury College, Vermont

Mr. Richard Lanier, Director, The Trust for Mutual Understanding, New York

Dr. Daniel C. Matuszewski, Director, IREX, Washington, D.C.

Professor Victor Myts, Director, Crimean branch of the Institute of Archaeology, Simferopol

Dr. L. M. Novokhatko, Deputy State Secretary, Ministry of Culture and Arts, Ukraine

Dr. Marilyn Perry, President, The Samuel H. Kress Foundation, New York

Ms. Nancy Pitman, The Brown Foundation, Houston

Dr. Jerry Podany, Conservator of Antiquities, J. Paul Getty Museum

Ms. Anne-Sophie Roure, Director, World Monument Watch, New York

Mr. Michael Sedoy, New York

Ms. Brooke Shearer, Washington, D.C.

Mr. John Stubbs, Vice President for Programs, World Monuments Watch, New York

Academician Petro Tolochko, Director, Institute of Archaeology, Kyiv

Mr. Daren Wilder, The James R. Dougherty, Jr., Foundation, Beeville, TX

Ms. Marta Pereyma, Assistant Cultural Affairs Officer, US Embassy, Kyiv

Region of Basilicata

Michele Brucoli, Dipartmento Agricoltura, Regione Basilicata

Filippo Bubbico, Presidente, Regione Basilicata

Gerardo Delfino, Ammininstratore Unico dell'Alsia

Raffaele DiNardo, Ex-Presidente, Regione Basilicata

Vito De Filippo, Vice Presidente , Regione Basilicata

Salvatore Martelli, Ex-Direttore dell'Ufficio Agricoltura, Regione Basilicata

Carmelo Mennone, Responsabile, Azienda Pantanello

Ms. Janet Demiray, Public Affairs Officer, US Embassy, Kyiv

Mr. Paul Hacker, Cultural Affairs Officer, US Embassy, Kyiv

Carmine Nigro, Assessore all'Agricoltura, Regione Basilicata

Sevastopol City Administration

Mr. Leonid Zhunko, Chairman of the Sevastopol City Administration

Mr. Andrei Miroshnichenko, Advisor to the Chairman of the Sevastopol City Administration

The University of Texas at Austin

Dr. Mark Yudoff, Chancellor

Dr. Larry Faulkner, President

Dr. Sheldon Ekland-Olson, Vice President and Provost

Dr. Juan Sanchez, Vice President for Research

Dr. Richard W. Lariviere, Dean, College of Liberal Arts

Prof. Fredrick R. Steiner, Dean, School of Architecture

Dr. Teresa Sullivan, Dean of Graduate Studies

Dr. Thomas Staley, Director, Harry Ransom Humanities Research Center

Prof. Timothy Moore, Chair, Department of Classics

Prof. James Brow, Chair, Department of Anthropology

Prof. Gregory W. Knapp, Chair, Department of Geography

Prof. Gil Rappaport, Chair, Slavic Languages

Prof. Tom Garza, Director, CREEES

Prof. John Clarke, Greek and Roman Art and Architecture, Art History Department

Prof. David Gracy, School of Information

Prof. Lawrence Speck, School of Architecture

Institute of Classical Archaeology

Staff

Austin

DIRECTOR
Professor Joseph Coleman Carter

ADMINISTRATOR
Carol K. Cook, MA

PUBLICATIONS EDITOR
Christopher Williams, BA

CHERSONESOS PUBLICATIONS AND TECHNICAL DEVELOPEMENT
Glenn Mack, MA

RESEARCH SCIENTISTS
Publications
Chora of Metaponto, Sanctuary
Professor Mariah Wade, UT Anthropology/ICA
Chora of Metaponto, Survey
Alberto Prieto, Ph.D. candidate, UT Classics
Dr. Stephen Thompson

Remote Sensing and GIS
Jessica Trelogan, MA (Chersonesos)

Geomorphology
Paul Lehman, Ph.D. candidate, UT Geography

DEVELOPMENT
Asele Surina, MA

CHERSONESOS COORDINATOR
Taissa Bushnell

OFFICE MANAGER
Pat Irwin

REMOTE SENSING
Prof. Melba Crawford, Co-principal investigator, NASA Project, Center for Space Research (CSR)

ARCHITECTURAL CONSULTANTS
Alma Maldonado Holiday, MArch., Llano Design, Austin
Carl Holiday, MArch., MS, Llano Design, Austin
Jodi Lane, MArch

RESEARCH ASSISTANTS
Gordon Lobay, Ph.D. candidate, Univ. of Cambridge
Jennifer Sacher, BA

TRANSLATION
Svetlana Telenkova, Kyiv (Russian & Ukrainian)
Roman Sazonov, BA (Russian)

COMPUTER SYSTEMS SUPPORT
Gary Lundquist

Centro di Agroarcheologia Pantanello, Metaponto

DIRECTOR
Professor Joseph Coleman Carter

ADMINISTRATOR
Dr. Rosetta Torraco

OPERATIONS
Cosimo D'Anzi, Cooperativa Pantanello (Maintenance of archaeological site)
Rocco Torraco, Azienda Pantanello (Technical support)

Pidtrymka Chersonesu, Sevastopol

DIRECTOR
Taissa Bushnell, MA

ADMINISTRATION
Lex Service, LTD